HUMAN
SUPERINTELLIGENCE

HUMAN SUPERINTELLIGENCE

How you can develop it using recursive self-improvement

John Stewart

The Glenroy Press

This book is for my mimetic and genetic grandchildren and great-grandchildren, especially **Winnie, Artie, William, and Jakey**

Table of Contents

Part 1: Introduction

1.

The Purpose of the Book

Are you able to see your thinking as 'object'? In other words, can you mentally stand outside your thinking so that you can examine it, evaluate it, and improve it where necessary?

Without this capacity, you cannot know in detail the thinking strategies that you use to solve problems and achieve your goals. Nor can you consciously and intentionally improve your thinking strategies in the light of experience.

Do you continually bootstrap your thinking strategies in this way? When you find that a thinking strategy fails, do you consciously and intentionally use this knowledge about the strategy to identify specifically where it is flawed and how it can be fixed?

If you did this actively and regularly throughout your life, you could continually enhance your cognitive capacities. By paying attention to your thought processes, you could identify where they could be improved and implement the relevant changes.

Do you regularly use such a meta-learning process and enhance it by actively seeking opportunities to test the effectiveness of your revised thinking strategies, amending them where they fail, and so on, repeatedly?

Potentially, you could also observe the meta-strategies that you use to develop and improve your thinking strategies. This would enable you to evaluate and improve them where necessary.

Are you aware in detail of the meta-thinking strategies that you use to generate your thinking strategies? If so, are you able to consciously and intentionally improve your meta-strategies when you find that they have produced a thinking strategy that is flawed? Do you take every opportunity to test your meta-strategies and to improve them continually? Do you then use your enhanced meta-strategies to improve not only other meta-strategies, but also themselves, recursively?

Do you celebrate when you find a flaw in your thinking strategies or in your meta-thinking strategies because it means that you have encountered an opportunity to improve your cognitive capacity?

Practices and associated knowledge exist that enable individuals to self-scaffold the ability to dis-embed from their existing thinking and meta-thinking, to see it as object, and to modify and improve their thinking as necessary.

Are you aware of any practices and knowledge that can be used in this way? If so, have you made serious and sustained efforts to use these practices? If you have yet to do so, but are persuaded in the future that these practices could work effectively, would you then make serious and persistent efforts to use them?

In recent years, additional tools have been developed that can enhance the process of recursively improving your own cognitive capacities. These tools identify the 'movements in thought' that constitute the thinking strategies that individuals with higher cognitive capacities use to build mental models of complex phenomena.

Otto Laske, an adult developmentalist, published a comprehensive description of these 'movements in thought' in his Manual of Dialectical Thought Forms.[1]

These tools have two invaluable uses for individuals who set out to scaffold their ability to build effective mental models of complex, dynamic phenomena:

First, the tools assist thinkers in identifying where their current thinking fails to represent relevant aspects of complex phenomena. The use of the thought forms helps thinkers see their existing thinking strategies as object and to see their limitations. This assists thinkers in identifying what thought processes are missing from their existing strategies and from the mental models they build.

Second, the tools assist thinkers in seeing what they need to include in their thinking and how they can include it if they are to

---

[1] Laske (2023) – see References for full citation

understand and manipulate complex phenomena. The tools guide thinkers in identifying how they can represent these aspects of complex reality in their models.

The 'movements in thought' are particularly effective for identifying and overcoming the limitations of what I refer to as analytical/rational thinking. This is the level of cognition that underpins mainstream science and that powered technological development, industrialization, and the First Enlightenment.

Analytical/rational thinking is very effective for understanding those restricted parts of reality that can be represented effectively by mechanistic, analysable, reductionist models. However, it is very limited in its ability to model and understand phenomena that are too complex and dynamic to be represented by mechanistic, analysable models. These limitations are the primary reason why mainstream science has failed to develop an adequate science of complex, dynamic phenomena. Unfortunately, this is a major restriction—most of the phenomena that constitute the reality we are embedded in are of this kind.

Are you aware of the discoveries made by research into the 'movements in thought' and 'thought forms' used by advanced thinkers to build effective models of complex phenomena? Are you aware of how individual thinkers can use these discoveries to scaffold their own ability to understand and manipulate complex reality?

If so, have you made efforts to practice the use of these 'movements in thought'? If not, but if you are persuaded in the future of the effectiveness of these tools, would you then be prepared to make the considerable efforts necessary to use the tools to scaffold the recursive self-improvement of your own cognition?

The practices that can be used to develop the capacity to dis-embed from thinking can also be adapted to scaffold the capacity to dis-embed from dysfunctional emotional predispositions, habitual behaviours, conditioning, and so on.

Are you aware of practices of this kind? If so, have you used them to free yourself from the dictates of dysfunctional predispositions and negative emotions? Can you see the advantages of doing so? Can you see even greater benefits in freeing yourself from all behavioural predispositions, enabling you to take whatever action is needed to achieve your goals in any given situation? If you knew that these kinds of practices can be effective, would you make serious and persistent efforts to use them?

Do you celebrate when you encounter difficulties and challenges, particularly ones that produce strong, unpleasant negative emotions, because this enables you to practice letting them arise and dissipate, so that they will not control your behaviour?

The adult developmental psychologist Robert Kegan produced a simple but powerful model of what occurs when an individual undergoes vertical development to a higher level in the cognitive or social/emotional domain. His model suggests that vertical development occurs when what was part of the subject at one level becomes object to a new, higher-level subject.[2]

He suggests that when an individual makes such a vertical transition in relation to their thinking, their thinking no longer has them. Instead, they have their thinking. And when they make this transition in relation to emotions, their emotions no longer have them. Instead, they have their emotions

In other words, the individual is no longer embedded in and controlled by their thinking or by their emotions. Instead, they can now choose if and when to engage in thinking and if and when to give attention to emotions that arise in their body. Thinking and emotions become tools that the individual can choose whether or not to use, freeing them to respond appropriately to the needs of whatever circumstances they are in.

During your life, have you ever set out consciously and intentionally to scaffold your own vertical development? Does Kegan's description of the fundamental shift that underpins vertical cognitive development resonate with anything in your own personal experience? What about the shift from emotions having you (e.g., being embedded in anger and being consumed by it) to you having your emotions (e.g.,

---

[2] Kegan (1982) – see References for full citation

noticing that anger has arisen in your body but choosing to let it go by without acting on it?)

The great Armenian systems thinker and developmentalist George Gurdjieff advocated the need for humans to become what he called 'self-evolving beings'.[3] Such a person would be able to remake themselves psychologically in whatever ways were necessary to meet the needs and demands of future evolution.

Their behaviour and psychology would no longer be controlled by the dictates of their biological and cultural evolutionary past and by their conditioning. Now, they would be able to move at right angles to the predispositions they have inherited and developed. Instead of serving the needs of past evolution, they would have the capacity to consciously and intentionally choose to do whatever is beneficial for the future evolutionary flourishing of humanity. As a self-evolving being, they would have the capacity to find motivation and satisfaction in whatever this requires them to do.

Are you aware of the concept of a self-evolving being? If so, are you aware of practices and associated knowledge that can be used to scaffold this capacity? Can you see how the emergence of self-evolving beings on any planet on which life emerges fundamentally changes the nature of the evolutionary process on that planet?

But what are the practical consequences of the emergence of self-evolving organisms on Earth? What will individuals who develop a capacity for self-evolution actually do with their lives? What strategies would they pursue if their goal is to contribute to the future evolutionary flourishing of humanity and life on Earth?

Fortunately, evolution science is beginning to answer this question as it develops a comprehensive understanding of the large-scale evolutionary processes that have made humans what we are and will shape our evolution into the future.

It is becoming increasingly evident that evolution has a trajectory. We can locate current humanity along this trajectory. This enables us to see how we must change our social systems and

---

[3] e.g., see Ouspensky (1949) – see References for full citation

psychology if we are to remain aligned with this trajectory in the future. If we can use our understanding of the direction of evolution in this way, we can survive and thrive indefinitely into the future, using the trajectory of evolution to guide what we do and how we evolve.

Broadly, evolution's trajectory is towards increasing integration and evolvability (evolvability is the ability to discover and implement effective adaptation. It improves as evolution advances. At the human level, a major component of this capacity is intelligence).

The trajectory towards increasing integration has produced the step-wise increases in the scale of cooperative organisation that characterise the long evolution of life on Earth. The first simple cells were cooperatives of self-producing molecular process. Cooperatives of these simple cells gave rise to the more complex eukaryote cell. Cooperatives of these produced multicellular organisms, and these eventually gave rise to cooperative animal societies.

This pattern has been repeated throughout human evolution, producing cooperative groups of increasing scale and evolvability: kin groups banded together to form tribal societies, cooperatives of these produced city-states, and further repetitions of this process produced kingdoms, empires, and eventually nation-states.

A proper understanding of this trajectory and the evolutionary forces that drive it reveals the next great step in the evolution of life on Earth: the integration of humanity and life on Earth into a unified, cooperative, and highly-evolvable global society.

It is worth emphasising at this point that this next great evolutionary transition in the evolution of life on earth is not something that will happen in the far-flung future. Rather, if this step does not occur this century, it will likely never happen on this planet. The destructive competition between nation-states, corporations, and other entities that is generating global environmental destruction and the threat of nuclear annihilation is highly likely to end human civilization this century unless significant changes are made.

Unless we take this next great evolutionary step soon, life on Earth will likely become a failed evolutionary experiment. Instead of hatching a global entity, life on Earth will be an egg that never hatches and instead goes rotten.

It is of utmost importance to recognise that the next great evolutionary steps towards increasing integration and evolvability will not occur automatically. If these steps are to occur, humans will need to drive them consciously and intentionally.

Natural selection cannot be relied upon to generate the emergence and development of an evolvable global entity. This is despite the fact that evolution has got us to this point automatically, driven by natural selection at the genetic and cultural levels. However, there is no population of global entities that will compete to produce the natural selection that could drive the evolution and development of global entities.

When the evolution of life on a planet reaches the point reached by life on Earth, it will proceed further only if it is driven consciously and intentionally. Humans, therefore, have a critically important role in ensuring that the evolutionary process on Earth continues successfully.

Viewing this unfolding evolutionary process from the broadest perspective, it is as if we are living within a process that is destined to hatch an integrated entity on the scale of the planet.

But this developmental trajectory has a surprising feature. Although evolution has proceeded automatically along the trajectory up until now, it will not continue to do so. Earth will fulfill its destiny of hatching a global-scale entity only if certain conditions are met: humans must wake up to the nature of the developmental process in which they are embedded; see that they have an essential role to play if the process is to be completed successfully; and decide consciously and intentionally to do whatever is necessary to establish and develop a highly-evolvable and cooperative global entity.

As I have noted, if humans do not do this, it is unlikely that human civilization will survive this century. Only by performing their apparent evolutionary function can humanity survive and thrive indefinitely into the future and contribute positively to the future evolution of life in the universe. It is only to the extent that an individual pursues pro-evolutionary goals that their life can make sense in a larger scheme of things which has the potential to continue long after they die.

The same evolutionary forces that drive this developmental process on Earth will arise on every planet where life emerges and evolves.

Are you aware of this emerging evolutionary worldview? If so, do you have the cognitive capacity to understand it in full and to work out in detail its implications for humanity in general, and for yourself in particular? Can you use it to identify how you can live a life that is meaningful in a larger, evolutionary scheme of things?

Does it seem plausible to you that you have a critical role to play in the future evolution of life on this planet and beyond? Have you grasped that if you are to undertake such a role successfully, you will need to develop the capacity to become a self-evolving being? Do you understand that this will be necessary if you are to free yourself from the dictates of past evolution so that you can find motivation and purpose in doing whatever is necessary to meet the demands of future evolution?

Is your current cognitive capacity sufficiently developed to understand that analytical/rational thinking is woefully inadequate for understanding the evolutionary worldview and its detailed implications, including understanding how a global society can be established and organised? Is it therefore clear to you that a priority for the advancement of the evolutionary process on Earth at this time is to spread across humanity the higher cognitive capacities needed to understand the relevant evolutionary processes and their implications? Do you also see that you face an evolutionary imperative to work on yourself to scaffold these higher cognitive capacities in yourself?

The central aim of this book is to facilitate the spread of higher cognition across humanity. Consequently, a key goal is to identify how we can recursively improve our own cognitive capacities and to scaffold higher cognition in others.

By higher cognition, I mean at least the capacity to construct mental models that underpin the ability to understand and influence complex, dynamic phenomena. Foremost amongst these are models of the large-scale evolutionary processes that have shaped us and will determine the future of humanity and life in the universe. This includes constructing models of the evolution of our social systems so that we can see how we will need to reorganise our societies as we evolve into the future.

The book's central focus on cognitive development serves a larger, wider goal. The book is a conscious and intentional attempt to facilitate the spread of the enabling capacities needed to empower humanity to advance the evolutionary process on Earth and contribute positively to the future evolution of life in the universe.

HUMAN SUPERINTELLIGENCE

2.

Recursive Self-Improvement and Superintelligence

The idea that 'superintelligence' can be achieved by recursive self-improvement was developed initially in relation to artificial intelligence, not humans.

According to leading futurists, philosophers and AI scientists, the eventual emergence of AI that is superintelligent is highly likely. It is only a question of when. But none of these thinkers have argued that superintelligent humans are at least as likely.[4]

The predictions about the inevitability of the emergence of superintelligent AI are founded on the idea of recursive self-improvement. Recursive self-improvement requires two interrelated capacities.

The first is the ability to self-improve intelligence. AI that has this capacity would be able to use its intelligence and creativity to re-make itself in ways that enhance its existing intelligence. For example, it could change its software or hardware in particular ways that increase it capacity to solve problems intelligently.

The second is the capacity that has the potential to turbo-charge the emergence of ever-improving intelligence. It involves the AI recursively using its ability to improve itself. First, it develops higher intelligence, then uses this higher intelligence to improve its intelligence further, then it uses this to improve further, and so on.

But would not recursive self-improvement apply equally to humans? What would prevent humans from developing the ability to use their intelligence to enhance their own intelligence?

---

[4] e.g., see Bostrom (2014) – see References for full citation

HUMAN SUPERINTELLIGENCE

In particular, humans appear to have the potential to become aware of their own thought processes, identify where the ability of those processes to solve particular problems could be improved, and adopt those improvements. Then, they could use their enhanced intelligence to discover and implement further improvements in intelligence, and so on and so on, recursively.

At least at first glance, the potential of recursive self-improvement to produce superintelligence in humans seems to be as feasible and conceivable as it is for AI. But if so, why have not the prophets of recursive self-improvement and superintelligence in AI also predicted the emergence of superintelligence in humans?

A plausible explanation is that they have no direct experience of recursively improving their intelligence. This is not to suggest that the futurists, philosophers, and AI scientists whom we are discussing lack intelligence. By any measure, some of them are amongst the finest minds around at the beginning of the 21st century. But their highly developed intelligence is not self-made. Nor do they appear to be in the process of re-making it to enhance it further.

Their writings support this interpretation. They do not claim to have intentionally and consciously improved their own intelligence. Nor do they point to others who have. More significantly, they do not even appear to understand how humans could actually escalate their intelligence. They have not developed adequate theories and mental models about how human intelligence functions, let alone used such models to identify how current levels of intelligence could be enhanced.

This view of the limitations of current futurists is reinforced by the fact that when they write about enhancing human capacities, they tend to focus on external technological aids, not on enhancing our internal cognitive processes.

I agree with the transhumanists that technological advances will inevitably enhance human capacities to some extent, as they have already. However, that is not what this book is about. Here, I will focus on the far more significant enhancements that can be achieved through recursive self-improvement of our internal psychological capacities.

12

The lack of understanding amongst futurists, philosophers, and AI researchers about the functioning of human intelligence and cognition is further evidenced by the current lack of knowledge about building human-level artificial general intelligence. It is widely accepted by leading thinkers that current AI, with its heavy reliance on deep neural networks, falls far short of achieving human-level general intelligence. This is despite the impressive ability of current AI to learn how to beat all humans at chess and other games and its ability to mimic some of the outputs of human cognition, such as engaging in apparently intelligent conversations.

However, this manifest lack of understanding about how higher intelligence can actually be produced does not seem to have shaken the belief that AI intelligence can escalate indefinitely. Far from it. In fact, this lack of understanding has seemingly removed any barriers to the blind extrapolation of improvements in AI intelligence way into the future.

If you do not have any idea about how higher intelligence can actually be produced, it is easy to imagine that AI can use recursive self-improvement to generate unimaginable levels of intelligence. Practical details and obstacles could only get in the way of such clear-sky hypothesising and dreaming.

However, fortunately for the prophesiers, not enough is known about superintelligence to refute their blind extrapolations and hand-waving about the potential of AI. This enables them to make their extrapolations with the same certainty that an intelligent person in the 19th century could blindly predict that there would be no limits to the speed that humans and their technology could achieve. Extrapolation of the progressive improvements that had already been achieved suggested that super speeds would eventually be reached, far surpassing the speed of light.

For these reasons, futurists can get away with making blind predictions about the potential of AI. However, it is not so easy to get away unchallenged with predictions about the potential of humans to escalate their own intelligence.

HUMAN SUPERINTELLIGENCE

Anyone who makes strong claims about such a potential will soon have to face an aggressive challenge: If you know so much about how humans can recursively self-improve their own intelligence, demonstrate that you have done so yourself. Show us that you not only can talk the talk, but that you can also walk the talk. Show us the money!

It is a similar reaction to that faced by anyone who starts telling people how they can achieve spiritual enlightenment.

Merely glancing at the title of this book tends to produce such a reaction: Seriously, who does this guy think he is? He seems to believe he is much smarter than Einstein. If he is superintelligent, why isn't he already world-famous? Where are his Noble Prizes? Why hasn't his scientific work been universally acclaimed by scientists? Why is he not a billionaire? How come he is not dating a famous actress or supermodel? Why have I not seen him on Joe Rogan's podcast? Or on the cover of Rolling Stone? And so on and so on, coloured by whatever fantasy they would hope to live out if they were a super genius.

You can be reasonably sure that anyone who is not scared off by these inevitable challenges will fall into either of two categories. First, they may be seriously deluded. Alternatively, they may be absolutely certain about how higher cognition can be scaffolded, and they may have used this knowledge to enhance their own cognition.

Which category do I fall into? Am I really claiming that I know how humans can use recursive self-improvement to escalate their intelligence significantly and that I will reveal this knowledge in this book? And if so, where is the evidence that I have applied this knowledge to myself?

These are reasonable questions to ask in the circumstances. However, unless you have developed higher cognition already, you are not yet equipped to judge the accuracy of my claims. In order to do so, you will first have to read this book in full so that you can develop an understanding of the methods it outlines for escalating your cognition.

But reading and understanding the book will not be enough. You will then need to make some progress in applying the practices and developing the skills that are outlined in the book. You will need to begin to work on recursively improving your own intelligence. The book will tell you how.

It is only after you make some progress in developing higher cognition are you will possess undeniable evidence that the practices outlined in the book are valid. And, it is only when you develop higher cognition to some degree that you will be capable of understanding the materials presented by the book.

However, in these introductory chapters, I will begin by attempting to give you a 'feel' for what it might be like to have higher intelligence and some of the challenges involved in trying to develop it. This will lead to some understanding of why individuals at current levels will not easily understand what higher intelligence entails. It will also explain why they cannot detect it easily in others. We will see that whatever your level of intelligence and cognition, it will not be obvious to you whether someone else is operating at a qualitatively higher level.

But first, I need to outline briefly what I mean when I use the terms intelligence and superintelligence. Broadly, in the sense I am using it, intelligence is the capacity to develop and implement strategies that enable the achievement of particular goals. You are using a form of intelligence when you figure out what you have to do to get what you want or when you solve problems.

Some forms of intelligence may be better than others in specific circumstances—they may be more effective at achieving particular goals. In this sense, it is possible to rank different forms of intelligence in terms of their effectiveness. Broadly, I will use the term superintelligence when referring to forms of intelligence that are far superior at achieving human goals than are current levels of human intelligence. As we will see, the way in which superintelligence achieves this is largely unimaginable to those at current levels of human intelligence.

Broadly, this usage of the concept of 'superintelligence' is similar to the way in which philosopher Nick Bostrom uses it. He defines it as "any intellect that greatly exceeds the cognitive performance of humans in virtually all domains of interest."[5]

It is worth emphasising here that the sense in which I am using the term intelligence includes the capacity to actually implement the strategies that are developed by intelligence. In order to act intelligently, individuals need to be able to implement the insights it provides. But many cannot do this at present, at least in some circumstances. This is because our strategies are often waylaid by conflicting wants and desires. To take a simple example, we may know that losing weight is in our overall interests, but we cannot find the motivation and drive to implement the strategies we know can achieve it.

For these reasons, if humans are to act intelligently, we need to be able to free ourselves from the dictates of conflicting motivations and predispositions inherited from our biological, cultural, and social past. Currently, these can sabotage the achievement of our larger goals. Knowing what to do is not much use if you cannot actually do it. How to develop such a capacity is a significant theme in this book.

Another important distinction is between improvements in intelligence that are horizontal and those that are vertical. Horizontal development applies the existing level of intelligence to new problems that can be solved at that level of intelligence. In contrast, vertical development involves the emergence of a new, higher level of intelligence that can solve problems that cannot be dealt with at the previous level of intelligence.

As an example, consider the level of intelligence that first became widespread with the unfolding of the European Enlightenment, which began in the 17th century. The new kind of thinking that spread at this time resulted from the emergence of a new, higher level of intelligence that powered the rise of science, technology, and modernity more generally. It entailed a vertical enhancement of intelligence. In contrast, the use of this new level of cognition to generate seemingly endless scientific and technological discoveries represents horizontal development.

---

[5] Bostrom (2014) – See References for full citation

It is also worth underlining here that vertical differences in intelligence and cognitive ability are not synonymous at all with differences in IQ.

This book will argue that the next great vertical development in human intelligence will drive the emergence of a global Second Enlightenment. The book will outline how this new level can be developed intentionally. The long sequence of scientific and other discoveries that will be produced by Second Enlightenment thinking will represent the horizontal development of this new level.

HUMAN SUPERINTELLIGENCE

3.

Recognising and Achieving Higher Cognition

This brings us back to the issues I raised earlier. How can you recognise this new, higher level of intelligence in others and in yourself? How will this new level of intelligence manifest? How can you develop it? What will acquiring Second Enlightenment thinking enable you and others to do that you cannot do now? More pointedly, how would you know whether someone who writes a book about how higher levels of intelligence can be developed knows what they are talking about? What will the money that they need to show you look like?

In order to begin to answer these questions, it is useful to examine some examples of vertical differences in intelligence that may be within your experience. In particular, I will focus on examples where you may have some direct experience of the operation of a lower level and also of the operation of a level that is higher than that. This will enable you to get some feeling for how the higher level might look from the perspective of the lower one, and vice versa.

The first example builds on an experience familiar to many owners of dogs and also to carers of two-year-old children. When a dog owner is annoyed by something the dog has done and raises their voice to chastise it, the dog may respond by rushing to the lounge room couch and pushing its head under it. The dog acts as if it believes that because it can no longer see its owner, its owner can no longer see it. The bemused owner stands behind the dog, whose rump is in full sight. The dog's devilishly clever strategy had failed completely.

Investigations of this kind of scenario suggest that the dog's tactics result from what has been termed 'slavery to its visual field'. An individual who is a slave to its visual field is unaware of anything outside its visual or other sensory fields. As the individual uses its intelligence in its attempts to solve adaptive challenges, it is incapable of taking into account anything outside those fields. It has no capacity to 'go offline' mentally from its visual field, to imagine what might be there in its external environment, and to adapt accordingly.

In contrast, most dog owners have a very highly developed capacity to go offline from their visual field and use mental representations that enable them to imagine how the world is likely to be when they are not actually seeing it. So, when owners mentally put themselves in the position of their dog, they tend to work on the assumption that the dog has a similar basic intelligence to themselves and are consequently astonished by the dog's apparent stupidity.

Children up to two or three years of age are also slaves to their sensory fields. They have not yet developed the capacity to go offline mentally and to take into account mental representations of what they are unable to sense. Consequently, parents often have a similar experience to dog owners when they are playing hide-and-seek with their toddlers. The toddler apparently believes it has successfully hidden from its parents by pushing its head under a pillow, although the rest of its body is in plain sight.

These examples illustrate a key way in which intelligences that are at different vertical levels differ from each other. Intelligence at the higher level considers factors to which the lower level is blind. When these factors are relevant to how the higher intelligence can achieve its goals, it can adapt far more effectively. But the lower intelligence cannot possibly understand why the higher intelligence acts the way it does, even if their goals are identical—the lower intelligence is incapable of considering these relevant factors when it assesses what it will do to achieve its goals. It cannot 'see' them.

When interacting with intelligence at a lower level, the higher-level intelligence can be completely mystified about why the lower level ignores important factors that are patently obvious and relevant. In contrast, when a lower-level intelligence interacts with a higher one, it tends to see the higher intelligence as acting in ways that are inexplicable to it and do not make sense. Using the criteria that the lower level takes into account when it adapts, the higher-level intelligence can seem deranged. Even where the higher intelligence is manifestly superior at achieving common goals, the lower intelligence will not understand how the higher intelligence does it—it will seem to be using magic.

In particular, in the case of the dog thrusting its head under the couch, there is nothing the bemused owner can say or do that will enable the dog to suddenly develop the capacity to represent mentally the owner standing right behind it. Its brain processes would have to be radically reorganised in order to develop such a capacity.

It is useful to look briefly at an example of humans differing vertically in intelligence. Adult humans generally have the cognitive capacity to go offline mentally, to form mental representations of relevant circumstances outside their visual field, and to use those representations to work out how they might adapt to achieve their goals.

But as we consider later in the book, there are a number of distinct levels of cognitive development amongst humans. These correspond to vertical differences in the capacity to construct and use mental representations and models. As we will see in detail, as individuals move up these levels, they develop the ability to construct mental models of increasing complexity. This enables them to adapt and strategize successfully in circumstances of greater and greater complexity.

Here, we will take a specific example that deals with the level of cognitive development known as the concrete operations level. At this level, individuals can envisage mental models of concrete features of their environment. These features are concrete in the sense that they can be seen, touched, and felt. Such individuals are able to build a mental model of some concrete features of their environment and envisage the possible concrete actions they could take to interact with these features in order to achieve their goals. In their heads, they can play with their mental representations, trying our different interactions in order to evaluate which sequences of possible actions will get them what they want.

The specific example I will examine concerns the ex-British Prime Minister, Maggie Thatcher. Her actions and statements indicate that she was highly capable and intelligent at her level, but was restricted primarily to the concrete-operations level. This enabled her to talk and behave in ways that seemed authentic to British voters who were cognitively limited in a similar way.

Famously, Thatcher declared with absolute conviction and certainty that: "there is no such thing as society. There are individual men and women, and there are families."

She was not engaging in political spin and manipulation when she said this. What she said accurately reflected her direct experience of the world. She was simply reporting how the world looked to her, as seen through the prism of her limited mental models. She was reporting precisely how reality is experienced by any human who is at the concrete operations level of cognitive development. She was as certain about her statement as a dog is certain that when its head is under the couch, its angry owner can no longer see it.

It is not difficult to understand why Thatcher and others at the concrete operations level fail to 'see' societies. To understand their perspective, we need to imagine experiencing the world as if we were restricted to concrete mental representations. When you walk through the streets of a major city and consider only the concrete events around you, you will never see a society. Nor will you see an economic system.

If you are restricted to the concrete level, you will see only concrete interactions between individuals. You will see individuals exchanging goods for money, people talking, and so on. In order to see beyond these concrete interactions and to see a social or economic system, you need to be capable of also developing mental representations that go beyond the concrete. You have to imagine processes and patterns that are not concretely experienced as you move through the world.

To 'see' a system, you have to construct mental representations of the abstract relationships between system components, including patterns of behaviour, feedback loops, and other system-level cybernetic processes. None of this will show up in your concrete experience. You must imagine them and hypothesise them.

Without these abstract mental representations, societies or other systems will not exist for you. They will not exist for you in the same way that a dog's owner cannot exist for a dog if the owner is not in the dog's visual field—the dog does not have the capacity to go offline and represent its owner mentally.

Just as a dog is a slave to its visual field, Maggie Thatcher and others at the concrete operations level are slaves to concrete thinking.

This does not mean that concrete thinkers are unable to talk intelligently about social systems and discuss their characteristics. But it does mean that the mental representations that they have in their heads when they do so are concrete. For them, the images they have in their minds when they talk and think about social systems are representations of concrete events, not abstract representations of systems.

As with the previous examples I have given, individuals at a higher level of cognition will often be completely mystified by the apparent inability of concrete thinkers to 'see' what is so obvious to them. They will often be astounded by the strategies developed by concrete thinkers to achieve particular goals—the strategies will often ignore factors that are critically important for accomplishing the goals.

In contrast, concrete thinkers will be equally mystified that the strategies developed by those at the next level are based on unfounded theoretical nonsense that has no solid basis in the real world they experience.

Because the world is often more complex than concrete thinkers can represent in their mental models, they will often find that their strategies are undermined by factors they are unaware of. They are frequently blindsided by events that 'came out of left field'. This is because their concrete thinking leaves out aspects of reality that are highly relevant to their strategies and goals. However, they tend to conclude that these events are intrinsically unknowable and cannot be understood by any form of rigorous cognition.

It is also worth noting here that the giant gulf between different levels of vertical development is not easily bridged. Vertical development typically requires the emergence of complex new brain processes. In the case of the dog, it is no small thing to install in it the mental capacity to go offline from its visual field and to build new mental representations of other relevant factors. It is much the same for Maggie Thatcher and others at the concrete level of cognition. There is no simple way to install in them the capacity to build abstract mental representations of systems that include new elements that are not concrete.

Significantly, this cannot be achieved by pointing out to them the limitations of their current thinking and by describing to them what they are missing. Or by setting it all out in a book and getting them to read it carefully. These approaches will not reorganise the minds of concrete thinkers in ways that build entirely new capacities. Instead, individuals at the concrete level will tend to interpret concretely any words you use to describe the factors left out by their current thinking.

When you wax eloquently about the characteristics of social systems that are highly relevant to achieving their goals but are absent from their understanding, they will interpret your words as referring to concrete events and interactions. The words that you use to describe these characteristics will not call up into their minds abstract representations of the characteristics. They will never have used or experienced the mental representations that your words and sentences are attempting to refer to. They cannot understand representations that are missing from their mental life. And they will be absolutely certain that how they 'do the world' is the right way, and that your way is deluded and wrong.

Typical concrete operations thinkers believe unshakeably that they can understand the answers to all questions that can be answered properly. When they cannot understand the answers that you are trying to give them, they can arrive at only one conclusion: you are talking nonsense.

It is similar for individuals who are at the analytical/rational level. As I have mentioned, this is the cognitive level that gave rise to the First Enlightenment and that produced modern science and technology. Analytical/rational thinking goes beyond concrete thinking in that it can include abstractions in the mental models it builds.

However, it tends to be reductionist, analytical, and mechanistic, which is ineffective for modelling complex phenomena.

Nevertheless, First Enlightenment thinkers tend to be absolutely convinced that their thinking is capable of answering all questions about reality that can be answered. They believe that they, as individuals, can understand all accurate answers. Like the concrete operations thinkers before them, they believe that models and understandings of reality that are produced by levels of cognition that are higher than theirs do not make sense and often border on woo-woo. They are slaves to analytical/rational thinking.

I will give one further illustration of the giant gap between different vertical levels of intelligence and of the difficulty that individuals at a lower level have in understanding how a higher level operates, including by failing to recognise the precise nature of its adaptive advantages. This example relates to vertical improvements in the capacity of the individual to actually implement the strategies that the individual has devised.

As mentioned, this line of development includes the capacity to free oneself from the dictates of motivations and emotions that would otherwise undermine the individual's ability to implement optimal strategies. This capacity is essential if we are to achieve our goals in the world. As we shall see, improvements in this capacity are critically important if humans are to deal effectively with the existential threats that humanity currently faces.

The lower level in this example is the dreaming state that all humans are familiar with. When we are dreaming, we experience ourselves as acting in ways that make sense while we are dreaming. However, our actions often seem absurd and irrational once we have awoken and remember the dream. If we examine our dream from the perspective of the waking state, we can gain some insight into what it is about dreaming that produces this apparently absurd and maladaptive behaviour.

In particular, we see that when we are dreaming, we have a very limited capacity to make conscious choices and consider alternative actions and perspectives. Our decisions tend to make sense in the dream because they are made using this limited and restricted awareness. But once we wake up from the dream, we see additional possibilities that we could have chosen. These alternatives did not exist for us in the dreaming state.

Consequently, in a dream, we often find ourselves acting automatically in response to narrow motivations and impulses. If we were facing similar circumstances when we were awake, we would have been able to 'stand outside' of the situation and consider alternative actions, at least to some extent. But in the dream, our awareness is embedded in these motivations and impulses and in the actions they drive. They seem to have absolute control over us. We have no conscious choice about them. In contrast to what we seem to experience in the waking state, there is no independent 'I' that makes choices about things.

To varying degrees, the world's great spiritual and religious traditions have pointed to a third state that is vertically above both the dreaming and normal waking states. This third state is often referred to as 'being awake' or as 'awakening in the midst of ordinary life'. It is claimed to be as different from the normal waking state as that state is from dreaming.

When individuals awaken in the midst of ordinary life, they typically see some of their previous behaviours and actions as absurd and maladaptive. When awakened individuals evaluate these past behaviours, they are able to take into account things that were previously not in their awareness. Importantly, they are now much more able to make conscious choices about whether to act on particular motivations and predispositions. Previously, these tended to be acted upon automatically, with little conscious evaluation. No longer are they largely embedded in and controlled by their motivational and emotional systems—instead of emotions and motivations having them, they now have emotions and motivations.

From a broader perspective, they are now free from the dictates of the predispositions implanted in them by their evolutionary past and their upbringing. In their awakened state, they can now consciously choose to move at right angles to these dictates whenever it is in their wider interests to do so. Their ability to achieve longer-term goals is no longer undermined and sabotaged by lower-level motivations, desires and drives that they were previously embedded in. They are well on the way to becoming self-evolving beings.

Once individuals awaken in the midst of ordinary life, they also have the capacity to dis-embed from their thinking. Thoughts no longer have them. Thinking is now something that they can choose to engage in consciously or not. They can be in the present moment with a still mind, rather than embedded in thoughts about the future and the past.

This enables them to access a still mind at will. This, in turn, is particularly conducive to using awareness, intuitions, and emotional intelligence that were previously crowded out and blocked by their almost incessant embeddedness in thinking. Dis-embedding from thinking also enables individuals to stand outside their thinking and see it as object in real-time. This enables them to see its limitations and see where they need to work on it in order to overcome its deficiencies.

However, dis-embedding from current modes of thinking will not automatically lift an individual to a higher level of cognitive intelligence. In order to achieve this, you will also need to develop the particular mental skills and thought processes that constitute cognitive intelligence at that higher level. Dis-embedding from your current level of thinking and feeling is an enabling step that is critically important, but it is not sufficient by itself.

As for any vertical shift in development, waking up in the midst of ordinary life is not simple. It is extremely challenging and requires significant persistent effort. It has been said that it requires the individual to work on themselves so hard that the soles of their feet will sweat.

Individuals in the normal, unawakened state experience reality and their being-in-the-world quite differently from those who have awakened in the midst of ordinary life. Until they awaken, they have not experienced what it is like to operate at the higher level, and therefore do not know the meaning of words that refer to those experiences. Consequently, any attempt to describe to them verbally or in writing what it is like to operate at that level will tend to fail. Furthermore, to them, the actions and behaviours of those who are awake will often seem absurd and inexplicable, warranting disparagement and even punishment.

The world's spiritual and contemplative traditions have developed practices that are intended to develop some capacity to wake up, at least while sitting in a quiet room on a meditation cushion. The difficulty involved in producing such a vertical transition is demonstrated by the fact that the traditions produce few individuals who are fully awake in the midst of ordinary life.

For the reasons we have been exploring, individuals who aspire to awaken find it notoriously difficult to identify teachers/gurus who are awake and who have the capacity to help them awaken. Furthermore, it is evident that the transition is rarely produced by reading a book or studying theories or explanations of awakening.

As we will explore in greater detail in the book, this is partly because vertical development invariably requires the development of procedural knowledge, not just declarative knowledge. Procedural knowledge is knowledge that is embodied in skills (including physical, emotional, and mental skills) and it is often indescribable using language. In contrast, declarative knowledge is knowledge that can be put into words.

A classic example of procedural knowledge at the physical level is the ability to hit an effective serve in tennis. You cannot learn to serve well just by reading a book about serving. It is the same at the emotional and cognitive levels. Philosopher Michael Polanyi was referring to procedural and other non-declarative knowledge when he said: "We know more than we can tell."

In fact, as we investigate vertical development more deeply, we will find that the next great vertical leap in human intelligence (the transition to Second Enlightenment cognition) requires the acquisition of both procedural and declarative knowledge. It necessitates integrating purely cognitive, thought-based abilities with capacities associated with emotions, feelings, and intuition.

More metaphorically, this can be referred to as the integration of the left and right brain, of heart and head, and of feminine as well as masculine ways of knowing. As we will see in detail, pattern recognition and other resources associated with these capacities that are not thought-based are essential for building the intricate mental models and representations that enable the understanding of complex phenomena.

I will refer to this higher-level Second Enlightenment thinking as metasystemic cognition. It is the ability to build and use mental models of complex, dynamic phenomena.

It is worth emphasising here that metasystemic cognition should not be confused with what is currently referred to as systems thinking or systems science. As will be discussed later in the book, most of what passes for systems science at present is actually an analytical/rational reduction of complex phenomena. It is what is produced when analytical/rational science attempts to understand complex, dynamic phenomena.

Metasystemic cognition is very different from what is contained in textbooks on systems thinking. Its models cannot be reduced to mechanistic diagrams. Genuine systems thinkers use metasystemic thinking to generate complex insights that are then reduced to produce the analytical/rational representations presented in the textbooks.

The significance of the transition to metasystemic cognition is also worth underlining here. It is much more important than other vertical transitions in cognition previously achieved by humanity. This is because most phenomena that impact humans significantly are complex phenomena. They cannot be properly understood and manipulated without metasystemic cognition. Without metasystemic cognition, you cannot make sense of and understand complex reality.

For example, it is only with the emergence of a capacity for metasystemic cognition that humanity can understand the large-scale evolutionary processes that have produced us and that will determine our future. Only then can we become aware of the trajectory of evolution and of the essential role that humans need to play in the future evolution of life on Earth. These realizations can provide meaning and purpose for human existence. They can achieve this without having to rely on belief in supernatural entities.

Furthermore, metasystemic cognition will enable humans to understand themselves psychologically. In turn, this will enable them to develop a comprehensive capacity to intentionally enhance their cognition and social/emotional capacities. These all involve complex phenomena that can only be understood and manipulated through metasystemic cognition. For the first time in human evolution, the emergence of metasystemic cognition will make possible a comprehensive capacity for the recursive self-improvement of cognition. It will enable us to remake ourselves consciously and intentionally in order to enhance our capacities to meet whatever demands we face in the future.

It will enable us to do the same in relation to our social and political systems.

Broadly, as we will see in more detail throughout the book, individuals who develop metasystemic cognition are superintelligent compared to those who are limited to First Enlightenment, analytical/rational thinking.

As I indicated earlier, I have spent time providing these diverse examples of vertical development in cognition in order to give you a feel for some of the central challenges you will face as you attempt to escalate your intelligence. My presentation of the examples was intended to go beyond limited analytical and reductionist representations of the relevant issues. With some luck, they may have evoked intuitions, feelings, and glimpses of relevant past experiences that helped to provide a deeper and broader understanding.

But it should go without saying that the examples and discussion I have provided so far do not demonstrate unequivocally the truth of my conclusions. They are not proofs that would be acceptable to First Enlightenment thinking. Rather, they have been chosen and structured to begin the process of facilitating your transition towards Second Enlightenment thinking and metasystemic cognition.

In summary, there is no simple, clear-cut way to enable you to understand clearly what is entailed in levels of intelligence that are higher than your own, to identify individuals who are at those levels already, to recognise individuals who are reliable guides for your transformation, and to specify unambiguously in words how you can achieve these levels. In short, attempts to show individuals how they can intentionally transition to a new level encounter a classic Catch-22—it is only when you are already at the higher level that you can understand what the new level entails and what you must do to achieve it. Being at the new level is a precondition for understanding fully what you must do to get there swiftly and directly.

Nevertheless, if you want to check me out before you read further, you can take a look at my book Evolution's Arrow, my papers published in international peer-reviewed science journals, and my other writings.

In the extremely unlikely event that you are already at the metasystemic level, you will see that my work could have been produced only by someone at that level. But if your cognitive center of gravity is at the level of an accomplished mainstream scientist, a respected IT specialist, a successful analytic philosopher, or some other exemplar of First Enlightenment thinking, it will not help you much. You will likely conclude that my work goes way beyond the evidence (that you can see) and is not analytically rigorous enough. You will conclude that it fails the tests of First Enlightenment science.

However, if you want to seriously consider the potential of humans to escalate their cognitive capacity indefinitely through recursive self-improvement, you will need to keep reading the book and begin to apply the practices and methods it identifies.

For all these reasons, the book is not written as a rigorous scientific treatise. If it were, it would not achieve its central goal of helping to escalate and spread higher cognition across humanity. It would fail to contribute significantly to the successful and rapid unfolding of the Second Enlightenment.

Instead, the book is structured and written in a way that attempts to evoke other ways of knowing that loosen attachment to thought-based cognition and enable its limitations to be overcome.

Story and narrative are forms of human communication that have proven capable of evoking other forms of knowing. They are the traditional, time-proven method of conveying understandings that cannot be adequately transmitted by rigorous, analytical exposition.

Accordingly, this book's central ideas are initially embedded in a cognitive autobiography, which constitutes Part 2.

It tells the story of how I stumbled upon methods that enhanced my own cognitive and social/emotional capacities using recursive self-improvement. It sets out to show how a rather odd individual who was emotionally stunted, autistically orientated, and otherwise limited in a number of ways (including to the extent of being physically, emotionally, and mentally handicapped in some respects), but who was perhaps gifted in some other ways, was able, due to accidents of genetics and birth, and also due to family circumstances that were often dysfunctional in their own particular way, and due to numerous cultural, social and environmental accidents, was able to discover ways to self-scaffold his capacities to higher levels, to know in detail how he did so, and therefore be able to help others to do the same.

As only story-telling can do, this form of writing is intended to evoke empathies, sympathies, intuitions, insights, past experiences, and emotions that lead readers to a deeper understanding of how they can use recursive self-improvement to enhance their own capacities.

Furthermore, a developmental narrative is particularly well suited to introducing in a staged fashion the sequence of steps that individuals need to take as they recursively improve their own cognitive capacities. Although the details will differ, the nature of the key steps undertaken by individuals will likely be similar. A sequential narrative is also suited to presenting the relevant concepts and experiences in a graduated fashion that makes them easier to understand, proceeding from the simple to the increasingly complex.

In Part 3 of the Book, I will distill the discoveries made during this odyssey into a detailed description of practices and associated knowledge that can be used systematically to develop higher cognition recursively.

If the cognitive narrative achieves its goals, it will enable the reader to understand sufficiently the words used in the declarative description presented in Part 3. This, in turn, will enable the reader to begin in earnest the recursive self-improvement of their own cognition and social/emotional capacities. It will equip the reader to begin implementing the relevant practices and begin to climb the mountain, slowly but inexorably.

In Part 2, I will begin my account of my developmental odyssey by recounting how I first stumbled upon recursive self-improvement as a method for scaffolding my cognition in 1968, when I was a 15-year-old boy at school in Brisbane, Australia.

HUMAN SUPERINTELLIGENCE

Part 2: A Developmental Odyssey

4.

Stumbling onto the Path of Recursive Self-Improvement

Getting 12 out of 20 on my first physics test in Year 11 at school was the trigger that launched me onto the path of recursive self-improvement. Up until then, I had found maths and science easy. Generally, I got 100 percent or close to it in maths and science exams throughout my schooling. Usually, I did not get full marks only when I made a silly mistake.

I had not done as well in other subjects such as English, History and Geography. This was particularly the case when examinations in these subjects started to focus on essays rather than the mere regurgitation of facts. Looking back, I was exceptional in any subject with clear and identifiable rules for generating the right answer, including where straight memorization was required. In contrast, in subjects like English, there was often no formula that I could discover and apply that would produce what was valued by the teachers. Unlike in maths and science, 'correctness' in these subjects seemed to me to be subjective to some extent. It was not always black and white.

It was only years later that I saw that these characteristics of mine were symptoms of an underlying condition.

But this reliable pattern was seriously disrupted by the first physics test in Year 11 in 1968. It was immediately obvious to me that the cause was associated with the introduction of a new kind of physics course into the Queensland education system that year. This new approach to physics teaching, known as PSSC physics, focused on imparting an understanding of the underlying principles and theories of physics. The course then tested the extent of a student's understanding by getting them to apply their knowledge to solving relevant physics problems. The proponents of PSSC physics proudly proclaimed that the rote learning of facts about physics would no longer be sufficient to do well in physics in years 11 and 12.

This shift in emphasis from memorization to understanding was reflected in the fact that in a number of education systems around the world, the PSSC exams were 'open book'. You could take the PSSC text book into the exam with you. This gave all students easy access to the relevant physics facts, but it would not necessarily get them far in the exam because of its focus on understanding. However, as it turned out, 'open book' exams were not implemented in our course, though they were for chemistry, which had a similar philosophy.

12 out of 20 was almost failing! What was I to do about it?

This serious challenge to my self-esteem brought to the forefront of my mind something I had read in a book a few years earlier. 'Psycho-Cybernetics' was the title of this book that my father had brought home from a library. The blurb on the cover suggested that it was some kind of bestseller. Key themes of the book were that an understanding of cybernetics and related scientific principles could be used to achieve success and happiness in one's life. I was particularly intrigued by the idea that as we grow and develop, we can intentionally change the characteristics we find in ourselves.

This approach resonated with similar themes in several books about the ideas of George Gurdjieff that my father had also brought me from a library. Gurdjieff was known as a spiritual master of sorts in the early part of the last century and also demonstrated a highly developed capacity for systems thinking.

A central theme of Gurdjieff's philosophy was that individuals can consciously work on themselves throughout their lives to develop self-mastery. We do not have to take ourselves as fixed and given, but can instead see ourselves as a work-in-progress. As we discover the existence of capacities and skills that would make us more effective, we can take steps to install them, continually remaking ourselves intentionally as we grow and develop.

However, as far as I could see from the books that I had access to, Gurdjieff did not provide much detail on how to work on yourself to enhance your capacities. But his key idea took root in my mind and predisposed me to be on the lookout for methods that I could use to enhance my capabilities.

What particularly caught my attention in Psycho-Cybernetics was its discussion about how some teachers in Russia were using ideas about algorithms to turbocharge their teaching methods. Instead of teaching facts about science and maths, or providing worked examples of how standard problems could be solved, they were teaching directly the algorithms that could be used to generate solutions to the problems. As I interpreted it, they were identifying abstract methods and strategies that could be used to solve whole classes of science problems and teaching these to their students.

My immediate reaction was: "Why aren't our teachers doing this already?" It would be so much more effective. And the long hours that I spent bored in class every day would be more stimulating. It would not be as good for me as pursuing my obsession with fishing, hunting, prospecting, and other outdoor adventures. But it would be preferable to sitting at the back of every class I attended, whispering to my mates about what we had heard on the radio the previous night and trying to avoid the teacher's gaze.

This failure of our teachers to implement a teaching practice that seemed to me obviously superior confirmed the poor opinion I held of them all, without exception. The school I attended (the all-boys Church of England Grammar School in Brisbane) was then reputed to be the top school in the state of Queensland in both academic and sporting achievements. A number of the senior teachers had written the textbooks that were used widely throughout the state. But all of them had physical and personality oddities of various kinds. The boys mercilessly exploited these. And they spent their lives teaching stuff that was generally obvious and boring.

My shocking result in the physics test brought me to an important realization. Waiting for my teachers to teach me the relevant problem-solving algorithms would be futile. They had no idea. Furthermore, there were no books available where I could read more about these approaches (this was 1968, computer science was in its infancy, cybernetics was little known publicly, there was no internet, and I was a somewhat odd 15-year-old boy living in Australia). I had to do it myself. I had to discover and develop the relevant algorithms on my own.

How was I to do this?

I decided to interview the top kids in the physics class in order to identify the problem-solving algorithms they used, and then install them in myself. A simple and powerful plan! What could go wrong?

I started to question the top students about the strategies they used to solve the problems that we had been set in our first physics test. However, I was surprised to find that they had very little conscious awareness of the methods they used. They could show me the formulas, calculations, and other workings they used to figure out the answer to a particular problem. However, they could not describe the abstract mental strategies that they had used to come up with the specific and concrete steps they took to solve a problem.

They could not identify any general problem-solving strategies they used, although they could set out the concrete steps they took, just like the teachers did in the worked-examples they used in their lessons. When questioned about why they took the particular approach they did in specific instances, they were unable to provide much insight.

It was not that they wanted to keep secret their problem-solving techniques. They were actually unaware of the strategies they used.

However, the interviews and interrogations were not completely useless. I picked up a few hints here and there.

My second strategy was to attempt to become aware of my own thought processes when I was confronted with a difficult physics problem. I intended to identify cases in which my thinking strategies worked and then reflect on why the strategies worked in those particular instances. My next step was to attempt to generalise the successful thought strategies as much as possible, in order to develop approaches that could be applied successfully to a wider range of problems.

When a strategy did not work in a particular instance, I would try to understand why it failed and then attempt to develop an improved strategy that did work. Whenever an amended strategy worked, I tried to identify why it did and how to improve it. Whenever I developed a more effective strategy, I would then try to generalise it further, and so on.

Eventually, I became increasingly able to reflect on the effectiveness of the meta-strategies that I was using to develop better problem-solving algorithms. As this occurred, I also began to work on systematically evaluating and improving the meta-algorithms. And so on, and so on, at all levels, recursively self-improving my problem-solving capabilities.

At first, I spent a lot of time looking inward in my attempts to identify the problem-solving strategies and thought processes that I was using. However, I did not see much initially. I could see why my interviews with the top students did not get me very far. But gradually, my problem-solving thought processes emerged out of the mist.

Increasingly, I became consciously aware of them, and could consciously amend them and try out different approaches mentally, in my head. The more I succeeded in making some of these thought processes conscious, and the more I started playing with ways to amend and improve them, the better I became at being aware of my problem-solving algorithms.

It is difficult to put into words how I experienced this developmental process. But at the time, I remember a metaphor rattling around in my head that somewhat captured the actual experience of what I was doing. It was as if I was reducing the size of the gap that unconscious intuition would have to spark across in order to solve a problem. My conscious development of problem-solving strategies was reducing the need for unconscious insights that were beyond my control.

Ultimately, my goal was to maximize the effectiveness of my conscious problem-solving algorithms and minimize or even eliminate the role of intuition and unconscious insight.

This approach produced positive results fairly quickly. Soon, I started to get marks that put me near the top of the physics class. And I achieved this with less work. Before long, I stopped doing homework insofar as it required me to write down the solutions to problems and show how I worked them out. Instead, I would restrict myself to looking at each problem set for homework, working out in my head an appropriate problem-solving strategy for it, checking mentally that it would work, and moving on if it did.

In the increasingly rare instances in which my accumulated problem-solving strategies failed, I would try to develop a new strategy, test it, and then amend my accumulated algorithms as necessary so that they could now solve the new kind of problem.

Homework was a breeze, and I did not need to pay much attention during lessons. I was a happy little chappy, sitting up the back of the physics class, talking covertly to my mates about things that really mattered to us in our lives then, not physics.

Soon, I extended the meta-thinking approach that I had developed for physics to all other subjects at school. Maths and chemistry did not put up much resistance. I also started doing better at English, but there were things required in English exams that were still largely impervious to my strategies. For English, I could not sufficiently reduce the size of the gap that the spark needed to cross. Intuition and insights were still needed to bridge the gap.

In particular, I was unable to devise algorithms that could fully produce the creative essays required for English exams. It was not until nearly forty years later that I realized that this could be done more or less comprehensively and that James Joyce and some other literary giants had, in fact, accomplished this to varying degrees. I discovered that James Joyce, in particular, seemed to have been propelled in this direction by the same underlying condition that probably played a major role in this aspect of my own cognitive development.

This left zoology. Unlike physics and chemistry, the teaching of zoology in Queensland had not yet been radically updated. The year 11 and 12 zoology courses did not even attempt to teach the problem-solving approaches and theoretical understandings that are at the center of most scientific advances in the real world. Instead, zoology still focused on the memorization of facts about zoology.

However, I had an exceptional memory. I also had plenty of spare time because I did not need to do much work on the other subjects and because my parents refused to buy a TV set because it would distract their children from homework and other things that were more important to my parents. So I swallowed the zoology textbook, and even now, I would still be able to pass a thorough exam about its contents easily. In addition, this memorisation process was greatly aided by my development of mental strategies that made memorization easier. Central to this was developing schema and connections that linked up the disparate facts, thereby significantly reducing the information content that needed to be memorized, and making it easier to retrieve relevant facts from memory.

But there was a price to pay for refraining from actually working out the answers to math, physics and chemistry problems. It hit home three years later when I was taking a university end-of-year physics exam. In examinations, I had to produce actual written workings for the problems I solved. I could not just check that I had the mental strategies that would solve the problem and then move on to the next one.

I froze in this exam when I found that I was having difficulty remembering some of the basic multiplication tables that I needed to use to work out the actual answers to the problems. This memory deficiency also extended to complex calculations such as long division. I was out of practice. I had hardly done any calculations for years.

I wondered why the examiners were asking physics questions that required complex computations when their job was to test whether students understood the relevant physics, not whether they could do basic maths. Surely?

But I got stuck on the details of a number of problems. I could not complete the calculations. My solution was to write down the procedures that I would use to solve each problem, step by step. But I did not (could not) actually apply those steps in practice. I explained why I did this in a note to the examiners in my exam paper. I wrote that I had demonstrated mastery of the relevant physics, if not of basic computation, and that surely that qualified me for full marks.

Apparently, they agreed and gave me a High Distinction for the exam. But after that, I practised my multiplication tables and other computational basics before each exam.

By the second half of year 11 and throughout year 12, I was ranked first in the school academically out of around 200 students. In the previous years, I was ranked around 15th.

In the final physics exam covering years 11 and 12, I got 92 percent, while the second student got 81 percent. Interestingly, both physics teachers also did the exam. Apparently, they wanted to provide some kind of independent test of the difficulty of the new PSSC course and of the exam they had set. The head physics teacher got 68 percent, and the other got 39 percent.

I mentioned earlier that initially, I was extremely disappointed with my teachers for failing to teach us problem-solving algorithms directly rather than taking us through worked examples and getting us to read the textbook. But their exam results underlined that their failure to do so was not due to laziness or ignorance. It was not their fault. They simply lacked the capability to do so.

It was not just that they lacked the ability to be conscious of their own problem-solving mental processes so that they could impart them directly to their students. As demonstrated by their results in the physics test, they also lacked effective problem-solving techniques. They were at least two vertical cognitive levels below where they needed to be if they were to teach algorithms directly and effectively.

Even if they had possessed highly developed problem-solving abilities, that would not have been enough. In order to teach students the algorithms that underlay their own abilities, they would need to be conscious of them. They would need to be able to 'see' their own problem-solving strategies as objects in their awareness.

Without this, they would have little capacity to describe to their students how they went about solving particular problems. Until they reached this level, their problem-solving methods and algorithms would not exist for them consciously and could not be used consciously to teach. Nor were the teachers capable of developing with their students a commonly-understood set of words and concepts that could be used to describe to students their problem-solving strategies.

The teachers were no more able to look inside their own minds and 'see' their own problem-solving algorithms than dogs are able to 'see' things outside their visual field, or Maggie Thatcher was able to 'see' a society and other large-scale systems, or First Enlightenment scientists were able to 'see' complex phenomena that cannot be captured by reductionist, mechanistic thinking. However, none of these individuals are responsible for their particular blindness.

At the time, I was not very charitable to my teachers. But they were acting in the only way they could, given their level of development. I now see that berating them for their failings would have been as senseless as abusing a dog for failing to be able to walk around upright on its hind legs.

Furthermore, if my teachers had been able to teach algorithms directly, I would have been deprived of a powerful motivation to look inside my own mind, develop the ability to see my own thought processes, and see how I could improve them recursively. I might never have been propelled along the path of recursive self-improvement.

Schooling in Queensland ended at the completion of year 12 with public examinations covering what had been studied in years 11 and 12. These public examinations were the gateway to university. Over 10,000 students sat for the exams in 1979, and I was ranked fourth overall. This was even though the ranking system compulsorily included the student's marks in English. As I indicated earlier, my meta-thinking techniques had not worked as well for English. On the scale of 1 to 7 that was used to indicate how well a student performed in individual subjects, I got a 6 in English and a 7 in each of the other subjects.

The events that surrounded my poor performance in a physics test at the beginning of year 11 launched me on a path of conscious, recursive self-improvement that has continued throughout my life.

Central to this path is the attitude that the individual is a work-in-progress that can be continually remade consciously and intentionally. Each enhancement of intelligence can be used to enhance further one's capacities, including by enhancing one's capacity to improve one's abilities, and so on.

My final years at school were just the beginning.

Try out this approach on some challenges that are of vital interest to you. Do it recursively. Extend it to other classes of challenges. And each time you encounter a problem that cannot be dealt with effectively, celebrate because it is an opportunity to discover how to adapt your strategies so that they can now solve wider classes of challenges.

You may not find it easy to make conscious what it is that you currently do intuitively and unconsciously, and then to enhance these processes consciously. However, as the book proceeds, I will set out specific techniques and practices that will enable you to scaffold these capacities.

5.

Attending the University of Queensland

Gradually, I began to apply my ability to recursively improve my problem-solving skills to domains beyond schoolwork. When I did so, it generally proved to be just as effective for developing better strategies as it had been for schoolwork.

However, I was not systematic or single-minded in applying my meta-skills to new areas. Often, it was more a case of doing it only after being frustrated several times in my attempts to achieve a particular goal. And often, when I did apply my meta-capacity, I used it partly consciously and partly intuitively. But most of the time, this was enough to get what I wanted. This minimalist approach saved time and energy. The full, conscious application of my methods took a lot of effort.

However, in retrospect, it is obvious that the goals I pursued at the time were often naïve and restricted by a lack of understanding of the wider world and broader human possibilities. I became better at developing strategies for achieving these goals, but the goals were often narrow in scope and extremely self-centered. I was 17 during my first year at the University of Queensland.

The central goal that drove me at the time was to go fishing.

I was obsessed with developing better techniques for catching fish. For years, I had been a voracious reader of Australian and American fishing and outdoors magazines. I harvested the fishing techniques they described, generalised them, and adapted them to generate new and improved methods for catching the fish that I had access to.

Another significant part of my obsession with fishing was that it took me outdoors to extraordinarily beautiful environments. For me, heaven was wading over a sand flat covered by crystal clear green water under a cloudless blue sky with the warm sun on my back, flicking out a bait for flathead and working it as if it were a soft-plastic, artificial lure across places that flathead were likely to lie in wait to ambush their prey.

When I left school, it was taken for granted that I would attend university. I was told repeatedly that as the top student at Queensland's leading private school, the world was at my feet. I could do anything. For parents, teachers, and other authority figures, this meant that I should do Law or Medicine.

But I had other ideas. I chose to enroll in a science degree, majoring in zoology. This was the closest thing that existed at the time to a degree in Marine Biology. The extremely limited thinking that underlay my choice was that I wanted to work outdoors and be able to spend as much time as possible fishing. It was a great plan, given my obsessions, tunnel vision, and limited understanding of myself and the world.

The long university vacations gave me plenty of opportunities to fish my way around Queensland. In my first year, I spent January on the surf beaches and rocky headlands of Stradbroke Island near Brisbane in South-East Queensland, May on the beaches, rocks, and jetties of Magnetic Island in northern Queensland, August on the beaches and creeks even further north on Hinchinbrook Island, and the next January fishing and skin diving on an uninhabited coral cay at the southern end of the Great Barrier Reef. At that time, university exams were held only at the end of the year, so I could get away with little work during the year, just cramming at the end.

My mates and I travelled cheaply, invariably camping for free wherever we went, including in city parks. But I needed some money. I earned this by writing fishing articles for national fishing magazines. When I was 17 and 18, during my first two years at university, I had half a dozen magazine articles published. I wrote one about each new exotic fishing location I went to.

I have mentioned that English was my poorest subject academically. I was far from being a natural writer. But in order to write good fishing articles, I studied the writing formats and approaches used by the best authors who wrote for Australian and American outdoors and fishing magazines. I abstracted the techniques and patterns they used and generalised them so that I could adapt them for my purposes. I did not write instinctively and spontaneously, drawing on unconscious, intuitive knowledge. Instead, my writing was explicitly guided by my conscious theories about writing and how it could be structured to achieve particular effects.

I can still remember very clearly the pattern and rhythm I used for the articles I wrote during this two-year period in 1970 and 71.

But things did not go so well for me in the Zoology Department at the University of Queensland. First-year zoology was largely a repeat of years 11 and 12 at school. Since I had already memorized the relevant textbook, I did not have to do any work to get a High Distinction in the exams.

However, as my intellectual horizons expanded, I increasingly realized during my first and second years that what was being taught in the Zoology Department was not science. Largely, it was just a collection of facts. We were not being taught the methods and techniques that science used to generate powerful new ideas and theories that could explain details about the world and make surprising new predictions.

Science taught by the Zoology Department was little better than basic school science. It required students to memorize and regurgitate the residue of facts and data that is left behind by genuine science as it moves forward creatively into the future.

My eventual clashes with the people running the Zoology Department were manifestations of a pattern that had existed in me for as long as I could remember. As I grew older, this pattern increasingly determined my behaviour. I had a problem with authority. Whenever I was under the control of people with greater power than me, I tended to feel suffocated and abused.

This visceral reaction grew as I became increasingly aware that people in authority often got it wrong. Given my strong drive to control myself and my environment, I felt very insecure when I was subject to the direction of relatively incompetent people who disrupted the plans and strategies that I used to maintain this control.

My mother was the dictator of my family. She controlled everything, including my father. I can remember that from a young age, I resented it when she used her power to prevent me from doing what I wanted. With my personality, I felt justified in doing whatever I could to escape her control on any issue. Given my characteristics and predispositions, one way I could do this was to develop my capacity to argue rationally and logically. I would use this to justify what I wanted to do, and to demonstrate that my mother's position was wrong.

This probably also incentivized me to develop arguments and theoretical positions that were more abstract and more encompassing. For example, I found myself using tactics to win arguments that involved re-framing the discussion within a bigger, wider perspective and by going 'meta'. My family environment drove me to become a master arguer at a very young age.

My distrust of authority and of most adults was reinforced when I started school as a five-year-old. My mother had decided that her two boys would never go to a public school. Since the Church of England Grammar School did not accept students until Grade 4, she first sent us to the local Catholic school. I quickly found that I was allergic to ritual, prayer, having to show reverence to any person or thing, and ideas about supernatural beings. I soon saw that the Mother Superior, nuns, and priests often acted in ways contrary to what they espoused and quoted from the Catechism and Bible. The Mother Superior, in particular, could be a very nasty piece of work.

My visceral antipathy toward religion was fuelled significantly when I eventually got to the Church of England Grammar School. I soon discovered that what they had to say about God, the Bible, and Christianity was substantially different from what the Catholics told me. They could not both be right. Increasingly, I found that both were wrong. It was manifestly the case that even those in authority who claimed to have God on their side could not be trusted.

However, for many years as I was growing up, I did not do much to confront those in power or try to overthrow them actively. There was no future in that. Instead, I coped using avoidance and manipulation.

This began to change in my final year at school. It was 1969, and student rebellion was in the air across the Westernised globe. It even penetrated Brisbane. Some of my friends at school started to get interested in these ideas and movements before I did. But when they began to agitate for setting up a school student council, they decided that it would be useful to have me as a member of the leadership group, given that I was the top student in the school.

I was a member of the first school council, and we set out to discuss changes to school policies with the Headmaster. Burning issues for us were abolishing the requirement to wear hats as part of our school uniform and the prohibition against long hair.

The Headmaster was highly regarded throughout Australia as a school principal. He was the head of the relevant national association of principals. Teachers and parents told us that he was a great man. But I soon found that he was evasive and inauthentic in his interactions with us. He was not interested in going where logic and rationality should have taken him on each issue.

Furthermore, to sustain the positions he adopted on issues, he was not above using his power and authority to stare down and bully students who continued to challenge him and point out the flaws in his arguments.

I was particularly unimpressed when he made me bend over his desk and then hit me vigorously six times with his blackboard ruler as punishment for not attending the compulsory chapel service that preceded classes each day. And he seemed to really enjoy doing it.

Even the Zoology Department at the University of Queensland began to be affected by student activism, although to a limited degree. Consistent with the strategies being pursued by activists across the university, I stood up in lectures and called for relevant classes to be cancelled to enable students to march in demonstrations against the Vietnam War.

Criticisms about how zoology and evolutionary science were taught in the university also grew. Eventually, the hierarchy could no longer ignore this. The head of the Zoology Department agreed to participate in a group of students and academic staff who would discuss proposals to change how things were being done. I was invited to attend as a prominent second-year student. The experience was underwhelming. It underlined and reinforced my views about the incompetence and dishonesty that seemed to characterise authority structures.

The first meeting lasted three hours. Only one issue was discussed but not resolved: whether the group should be referred to as 'The Zoology Group' or 'A Zoology Group'. Apparently, the hierarchy was concerned that calling it 'The' Zoology Group would have wrongly conveyed the impression that it had some official status and power within the university.

The meetings fizzled out, achieving nothing of significance.

In large part, my disaffection with the way the Zoology Department went about its teaching was because I knew they could have explored perspectives that were far more interesting and stimulating. This was particularly the case in the areas that interested me the most: evolutionary science and ecology.

I had been exposed to 'big picture' and systems perspectives about evolution and ecology in books that my father brought home from libraries when I was in my early teens. My father worked in a bank from the age of 18 until he retired at 63. But his mind did not spend much time on his bank work.

In his head, he was a philosopher, vitally interested in trying to understand the meaning and purpose of human existence, including his own. He was a member of the Brisbane branch of the Theosophical Society and plundered its library in his quest for new ideas and perspectives. The Theosophists tended to be more serious and intellectual than the New Agers, but they went way beyond the science of the time in their quest to understand the human condition.

My father enjoyed having a son who seemed smart enough and interested enough to pursue these kinds of ideas throughout his life. He brought home a wide range of books. Psycho-Cybernetics and Gurdjieff books were not the only ones that eventually deeply impacted me.

I did not fully realize until many years later what an extraordinary range of cutting-edge books he had gotten for me to read.

Unfortunately, it was too late then to thank him and express how much the books had contributed to making me who I became, for better and worse.

Key books he brought me included Karl Popper's treatise on the philosophy of science, 'The Logic of Scientific Discovery', Teilhard de Chardin's evolutionary epic 'The Phenomenon of Man', Madame Blavatsky's 'The Secret Doctrine', Heidegger's 'Being and Time', Alfred North Whitehead's 'Science and Society', and many others.

In general, I enjoyed reading these books, but they did not have a strong and immediate impact on me. I could generally follow their ideas, found them interesting, and thought that much of what they had to say made sense and was even obvious. The exceptions were when they went off into what I saw as religious and supernatural wishful thinking. Teilhard de Chardin, for example, tended to do this near the end of his books.

However, I was not put off by speculative thinking seemed to be more grounded in material reality. I learned from Karl Popper that speculation and imagination served important functions in the process of scientific discovery. They were essential for generating bold, innovative hypotheses that could then be subjected to rigorous testing. As Popper established, the process of 'conjecture and refutation' is central to the advancement of science. As he emphasised, coming up with bold, innovative, and potentially falsifiable conjectures is at least as important for the success of science as is the collection of relevant evidence and data.

I was fully expecting that when I went to university, I would start learning real science. Physics and chemistry in the final two years of school had begun to move in this direction. Zoology had not, but I looked forward to this changing at university. Disappointingly, the first zoology lectures were still all about facts and memorization. But being somewhat naive and optimistic at the time, I expected that very soon, there would be a lecture that began to teach us about Teilhard de Chardin's grand vision of evolution. I hoped that before long, we would discuss the deep implications of large-scale evolutionary processes for humanity and our societies.

But it never happened. There was no mention of de Chardin's ideas in any lecture I attended during my science degree. And I did every available course that had anything to do with evolution.

Almost without exception, the academic staff in the Zoology Department seemed to be focused only on data collection—measuring, counting, and classifying. Some of them openly disparaged the role of theory in evolutionary science. The deputy head of the Department told me that all theory was ephemeral and that only facts represented unchanging truth.

I responded that without good theorizing and hypothesising, researchers would not know what data to collect in order to advance scientific knowledge. An infinite amount of data could be collected in the world, but only a tiny part of it is relevant to testing new ideas and thereby contributing to the progress of science. Yes, if a researcher counts all the grains of sand on a beach, the outcome would provide a fact about the circumstances at the time, but it would not be relevant to anything that matters. And so on and so on, I argued, drawing on a Popperian conception of science.

But I doubt seriously whether he had read and thought deeply about key issues in the philosophy of science, let alone understood them. I did not encounter any members of the lecturing staff who had read Popper and who could sensibly discuss his ideas and their relevance to their work. An intellectual wasteland.

I used to say that you could fire a rifle through the middle of the Zoology Department and miss the nearest new idea by three kilometers. I also suggested that the Department failed to meet the basic requirements for being part of a university science faculty. It was not doing or teaching real science. Instead, it was doing only technical work and training data collectors and technicians. Consequently, I argued, it should be excised from the University of Queensland and instead incorporated into the Queensland Institute of Technology.

It amused me at the time to say these kinds of things to authority, but the joke was on me a decade or two later when the Queensland Institute of Technology itself was given full university status.

6.

Going Fishing and Finding Philosophy

In the final analysis, my struggle between the conflicting motivations of continuing at university or going fishing could end in only one way. There was no real contest. I remember the day towards the end of my second year at Queensland University when I decided to resolve this conflict. Sitting in the Zoology Department library, I came to the realization that instead of pursuing a career path in Marine Biology that would give me some opportunities to go fishing, I should leave university immediately and go fishing full time.

This decision was made even easier because I had discovered during my trips to Northern Queensland that it was possible to earn a living fishing for Spanish Mackerel and reef fish on the Great Barrier Reef. Importantly for me, this involved catching the fish one at a time on fishing lines, not using nets. This form of fishing had all the key elements that drove my fishing obsession. The level of income of a professional mackerel fisherman depends on the choice of baits, lures, lines, locations, and so on. The better the techniques that the fisherman develops and adopts, the better he will do.

That day, I decided to take at least four years off university to become a professional mackerel fisherman on the Great Barrier Reef. The plan was to get a job in northern Queensland in order to earn enough money to buy a mackerel fishing boat that I could live on. Given that I was fully equipped with masterful problem-solving algorithms, how could this plan fail to succeed? These included algorithms for developing new and improved algorithms as needed. With these tools, I was exceptionally good at working out how to get what I wanted.

Early in the morning, a few weeks later, with 28 dollars in my pocket, I got my mother to drop me off on Highway One on the northern outskirts of Brisbane. I began hitch-hiking to Cairns in Far North Queensland. I told my mother that it was just another vacation fishing trip I was embarking on and that I would return in a few months to continue at university. I wrote to my parents from Cairns a couple of weeks later to tell them that I would not be back anytime soon and would not be starting third-year zoology.

I had left behind family, friends, and most of my history. It felt exhilarating to cut all ties. Free at last! I loved the smell of burning bridges in the morning.

Seven or so months later, I was living the dream. I had spent six months working in Weipa, an isolated bauxite mining town on the western coast of Cape York Peninsula. I worked in railroad construction, assisting with welding railway tracks together. The wages and conditions were exceptionally high, given the need to attract a temporary workforce prepared to be separated from families and friends, work in a town with few amenities, and live in very basic temporary accommodation.

I quickly accumulated the money to return to Cairns to buy a 26-foot licensed mackerel fishing boat that I could live on. The federal government duly issued me with a Commonwealth Master Fisherman's Licence. I was soon heading out of Cairns to the Barrier Reef for up to five or six days at a time, out of sight of land and catching fish on reefs that were tens of kilometers north and south of Cairns. Being an introvert and a loner, I generally went out alone. I had never owned a boat before, and had never taken one out to sea by myself. I had some near misses. At 19 and with zero relevant experience, there should have been a law against it.

But I survived and loved what I was doing. After a few months, my plans broadened. The idea now was to spend a few seasons fishing out of Cairns, then upgrade to a bigger boat, and head to the Torres Strait and to the islands of the western Pacific outside the Barrier Reef. Apart from the fishing and the beauty of reefs and islands, part of the attraction was island girls. In my experience, many were spontaneous, loved life, and had a great sense of humour. And they were just like Margaret Mead said they were.

But other events intervened, eventually propelling my life down a different path.

The immediate cause was the Southeast Trade Winds which regularly battered the Barrier Reef and the northeastern coast of Queensland. Often, the trade winds blew consistently for a week or more, preventing mackerel fishermen from going to sea and catching fish.

Consequently, I had to find something to do while I waited for the winds to drop. As a reader and a loner, I gravitated to spending my days in the Cairns Library. The library was a classical two-story colonial-style building with large cool verandas in the center of Cairns. I would happily spend full days in the library reading newspapers and books.

Since I always liked to have a system to guide me, I decided to begin reading the books that were first in line on the library's shelves, and then work my way along from there. As luck would have it, the library classified its books using the Dewey system. My strategy quickly led me to books on philosophy and psychology.

Soon, I was hooked. The books I read on philosophy resonated strongly with the ideas I had gotten from the books my father had brought home for me when I was a young teenager. In contrast to university science which had held little interest for me, the philosophy I read was full of stimulating ideas and 'big picture' perspectives. It regularly went to the 'meta' level, focusing not just on particular ideas, but on ideas and perspectives about those ideas.

Of particular interest to me was that I found several philosophical books that dealt with questions and issues that really mattered to human beings, including me. Where do we come from? What are we? Where are we going? And most significantly, what should we do? How should we live our lives? How should we interact with others? What, if any, ethical principles should we use to guide us through our lives?

Despite the enormous gulf between the philosophy I was now reading and the comparatively 'brain dead' evolutionary science that filled my university zoology course, my intellectual awakening in the

Cairns Library did not turn me away from the evolutionary perspective. Rather, it further fuelled my deep intuition that understanding the evolutionary processes that have shaped humanity was the key to answering the existential questions we face. I intuited that academic evolutionary science and thinkers like Teilhard de Chardin and Desmond Morris had barely scratched the surface of this potential. But at that time, I was not able to come up with a viable alternative set of theories. Nevertheless, the seed had been planted that was to govern my intellectual development for the rest of my life.

Furthermore, my reading about philosophy convinced me that I was not being overambitious in believing that I could develop a coherent and relevant evolutionary theory and philosophy. With my recursively self-improving mental tool kit, I found that I had no trouble following and understanding the ideas developed by the great philosophers, identifying deficiencies in their arguments, and working out how those limitations could be overcome.

I remember being very impressed with the work of the great English philosopher Bertrand Russell. However, I had little doubt that I could do the kind of work he had done. It was not until years later that I saw that even though he was obviously a genius of sorts, he was a genius only at the level of analytical/rational thinking. Eventually, I realized that if I were to succeed with the development of my evolutionary philosophy, I would need to develop cognitive capacities that were vertically above those of the hero of my younger years, Bertrand Russell.

While in the library, I noticed that I was not the only person spending hours a day there, completely absorbed in reading and thinking. A number of old, bearded, dishevelled, dirty, apparently homeless men were often in the library, focused on reading the daily newspapers. Some would furiously and triumphantly underline particular passages they found in the papers. They acted as if they had just discovered and confirmed special knowledge that was highly significant but that only they could see. Sometimes, when I noticed this, I would chuckle internally and wonder whether it was my destiny to become one of them. Fifty years have passed, and I have not ended up like that. Yet. But there is still time.

The realization that I wanted to spend my life working on evolutionary philosophy changed everything. Doing so was completely incompatible with being a professional fisherman who wandered nomadically around the Barrier Reef and the Pacific Islands in his boat.

I had already discovered that heavy physical work tended to prevent sustained and deep intellectual work. During my heavy labouring work in railroad construction in Weipa, I was physically exhausted at the end of each 10-hour work day. And although I loved it, fishing was even more of a barrier. Living and working in a small boat fifty or more miles off the coast, navigating through dense reefs and continually contending with challenging sea conditions, was a very immersive experience. A professional fisherman in a little boat was continually absorbed in their senses and being physically tested to their limits. As the Torres Strait Islanders used to say, it involved spending your days 'chewing the salt water'.

I decided that I needed to find work that was not physically challenging in a city with good libraries that was located away from distracting fishing opportunities. With this in mind, I sold my fishing boat at the end of the year and headed south, leaving my beloved tropics and the Great Barrier Reef far behind.

Eventually, I ended up in Canberra, Australia's national capital. It is a planned city established in 1913 by the federal government on what was formerly vacant bushland. Canberra is a government town with great public facilities, including libraries and the Australian National University.

I started employment in the federal government in 1972 as a 'Clerk Class 1', the lowest clerical and administrative work level. Initially, I had plenty of time for reading and thinking. However, I got promoted up the scale quickly, which required more demanding work.

And there were other distractions. Having attended an all-boys school, I had not learned much about interacting with females by the time I went to university. In Canberra, I formed my first longer-term relationships with girls, lived with one, and was strongly attached several times. I found that I very much enjoyed female company. It was not just Island girls. I discovered that it tended to give me some relief from continually thinking about how to control my environment and get what I wanted.

I had two particularly interesting jobs during my time in Canberra. One was as an administrative and research officer with the Priorities Review Staff. It was an elite think tank set up by the government to advise it about the economic, social, and other policies that it would need to achieve its longer-term societal goals. This think tank was to focus on the future rather than immediate needs.

The other job was as a policy and admin officer with the International Women's Year Secretariat. Its function was to advise the government on policies that would support 1975's International Women's Year. As I have already indicated, I was a beginner at understanding women's issues. However, I had embraced feminism to the extent possible given my limited background. I was particularly influenced by Germaine Greer's 'The Female Eunuch', and read other books and articles. I began to understand the evolutionary, cultural, and social forces that often stood in the way of women achieving their goals in society, individually and collectively. I was more than happy to be a token male within the Secretariat.

After two years working for the government in Canberra, I resigned from my Class 6 position and headed north again for the tropics and the Great Barrier Reef. I was enjoying my life in Canberra, but I had one year left before my planned four-year break from university ended. I decided that before I returned to finish my degree, I would get another fishing boat and spend a final season fishing for mackerel on the Great Barrier Reef in Far North Queensland.

7.

Back to University

I returned to Brisbane and Queensland University at the beginning of 1976 as a 23-year-old. I intended to complete my degree and then pursue a career in academia as a theoretical evolutionary scientist.

I soon found that the Zoology Department had not changed much during my absence. But there were some improvements. The third-year subjects now offered in evolutionary science and ecology were considerably more interesting and stimulating than those available in earlier years.

Some lectures covered the latest developments in evolutionary science, including Edward O Wilson's 1975 book 'Sociobiology: The New Synthesis'. Controversially, Wilson applied some of the key findings made by evolutionary science about the evolution of sociality amongst animals to human evolution.

However, even these lectures still fell far short of what I intuited was possible for a 'big picture' evolutionary theory that applied to all aspects of human evolution, including the emergence of our political and economic systems. At that time, I could not see even the broad outlines of such a unified theory of evolution. But I knew that Wilson's work, admirable as it was, was just a small, stumbling step in that direction.

I did not understand it then, but I realized later that I would need to undergo further vertical transitions in my cognitive capacities if I were to develop such a new, comprehensive theoretical understanding. And it took much longer to realize that any such theory could be properly understood only by others who had undergone similar vertical cognitive transitions.

I made some good friends in the final year of my degree. These were fellow students who lived and breathed zoology. Their enthusiasm was almost infectious. However, I realized that my interests and intellectual predispositions were very different from theirs. They were as passionate about bird watching and about collecting and identifying frogs as I was about fishing. But they were not at all theoretically orientated. They did not share my passion for developing an overarching understanding of evolution that could contribute significantly to understanding the human condition, including by answering the big existential questions.

I met one of these fellow students many years later, and reminisced with him about our time together at university. He told me that what startled him about my evolutionary thinking at the time was that it abstracted away 'the whole of nature'. For him, my emerging theories and ideas were so abstract that they did not include the concrete details of actual living processes in nature that he and our other friends were so passionate about studying and exploring.

But it was not just some of my fellow students who were motivated by passions that differed greatly from mine. I soon found that none of the academic staff in the Zoology Department shared my interests. I saw that if I were to pursue an academic career in evolutionary science at the University of Queensland, my PhD would have to focus on data collection. I would have to spend years counting and measuring animals.

The typical PhD completed in the Zoology Department was about 95 percent counting and measuring and about 5 percent theory. Ideally, I wanted to do a PhD that was 100 percent theory. I accepted that counting and measuring were an essential part of the scientific process, but I was more than happy to leave that task to other scientists. I wanted to specialize in what I was exceptional at.

As these realizations hit home, I set out to explore the possibility of pursuing an academic career at some other university in the world that valued theory more highly. I progressively familiarized myself with the latest journal articles in evolutionary science and related fields and identified the institutions and researchers that generated them.

This gave me a great overview of the state of evolutionary science at the time, and I found the survey enjoyable and stimulating.

Ultimately, however, I was very disappointed. I discovered that the state of evolutionary science at the University of Queensland was fairly typical, not unique. 'Big picture' evolutionary theorizing seemed to be frowned upon almost universally.

A number of scattered groups used mathematical modelling to address some theoretical issues. However, standard mathematical models cannot effectively deal with complex, organic systems. They are useful only for representing relatively mechanistic systems that can be analysed into components that interact in more-or-less fixed ways.

Standard mathematical modelling can deal effectively only with phenomena that are analytically tractable. Unfortunately, most complex evolutionary phenomena are not. Over-reliance on mathematical modelling and analysis reflects a key difference between First Enlightenment science and Second Enlightenment or metasystemic science. The use of complex simulations does not, by itself, enable this vertical gap to be overcome. Analytical/rational thinking has as much difficulty understanding complex simulations as it has understanding other complex phenomena—metasystemic cognition is essential in both cases.

My research showed that 'big picture' evolutionary science was 'on the nose' across the planet. For an evolutionary scientist to focus on these issues was career limiting. The only time it was safe for researchers to present 'big picture' evolutionary perspectives was when they wrote the final book of their careers. Typically, in the last chapter of their last book, they would outline their views about the implications for the human condition of the large-scale evolutionary processes that have shaped humanity and that will govern our future.

It was not until many years later that I discovered the central historical reason for evolutionary science's restrictions and limitations. It was no accident.

Since the middle of the last century, mainstream evolutionary science has largely accepted a set of ideas about evolution that became known as the Modern Evolutionary Synthesis. In general, the Synthesis rejects the existence of large-scale directionality in evolution and other patterns identified by 'big picture' evolutionary thinking.

But significantly, excluding these ideas from the synthesis was not due to any scientific case against large-scale directionality in evolution. In fact, these exclusions have an entirely different explanation. The contents of the Modern Evolutionary Synthesis were largely shaped by a small group of leading evolutionary scientists during the mid-1940s. They set out intentionally to fashion a particular version of evolutionary science that could survive and thrive as an academic discipline. To achieve this, it had to be shaped in such a way that it could avoid controversies and criticisms that might otherwise have retarded its acceptance from the very beginning.

As evolutionary philosopher Michael Ruse details in his book 'Monad to Man', the founders of the Modern Evolutionary Synthesis got together several times during the 1940s to plan how their proposed Synthesis could be established. Ruse indicates that a central concern of these meetings was to entrench evolutionary studies as an accepted academic discipline within science.

With profession-building as a key goal, the architects of the Synthesis decided that ideas about direction and progress should be excluded from the discipline of evolutionary science. In particular, the founders were concerned that some ideologies might attempt to manipulate such ideas to justify racism and social inequality. At the time, Hitler's misuse of evolutionary ideas had demonstrated the reality of this dangerous potential. Leaving this door open could embroil evolutionary science in controversy and undermine the academic acceptability of the nascent discipline.

Ruse outlines how the founders implemented their plan. They decided to use their editorial positions with scientific journals and their powers as 'respected' peer reviewers to influence what was publishable. In a world of publish or perish, these were powerful weapons for achieving their goals.

As Ruse points out, it is ironic that the architects made these decisions despite most of them supporting the view that directionality in some form or another is evident in evolution. See pages 438 to 450 of 'Monad to Man' for more detail. In particular, pages 447 to 450 focus on how directionality and wider 'speculation' were excluded from the Synthesis.

Against this background, it is understandable that new generations of evolutionary scientists came to believe that the exclusion from evolutionary science of the study of directionality and other large-scale patterns is because there is no scientific support for these ideas.

However, as Ruse makes clear, the initial exclusion was not due to relevant scientific considerations. There never was and never has been any convincing science-based justification for their exclusion. The exclusion was decided without any recourse to the widely accepted criteria for distinguishing speculations that are acceptable within science from speculations that should properly be excluded from serious science—i.e., the Popperian criteria of testability and falsifiability.

The decisions and actions of the architects of the Synthesis gave birth to the 'dark ages' of evolutionary science. We are only beginning to emerge from this era today. It has tended to produce the kind of science that results whenever reductionist, mechanistic science alone is deployed to understand complex, evolving phenomena. In the main, it produces rigorously justified trivia, precise and analytically rigorous mathematization that bears little relationship to reality, masses of highly accurate data that decides nothing of importance, an inability to say much of interest about large-scale complex phenomena, and journals full of junk papers that are soon forgotten.

To date, evolutionary science has failed miserably to fulfill its enormous potential to contribute significantly to humanity's survival and future evolutionary success. Evolutionary science put a gun to its head in the 1940s and blew its frontal lobes away.

But my intention here is not to deride the architects of the Modern Evolutionary Synthesis. They were reacting rationally and pragmatically to the circumstances that prevailed when they met to develop a synthesis. They had good reason to be extremely wary of the potential for evolutionary ideas to be misused. The Second World War was in part fuelled by racist ideologies that some had attempted to justify with quasi-evolutionary perspectives. Furthermore, if evolutionary science was to be accepted by mainstream science as a serious and legitimate academic discipline, it had to conform to mainstream views about what constituted 'proper' science.

As mentioned previously, mainstream science then and now is underpinned by First Enlightenment, analytical/rational thinking. This level of cognition is very good at understanding mechanistic systems that are analysable. However, it is inadequate for understanding complex organic phenomena. The stunted form of evolutionary science that the founders of the Modern Synthesis established was the only kind that was likely to be acceptable to the wider scientific community at the time. The advent of an evolutionary science that is able to model and understand large-scale, evolving, complex evolutionary processes is still awaiting the emergence of Second Enlightenment science and metasystemic cognition.

The evolutionary science that I surveyed in my final year at the University of Queensland continued to be retarded by profession-building decisions taken by the architects of the Modern Synthesis thirty years before. I had no ambition to be part of it.

What was I to do, given my strengthening intuition that a comprehensive understanding of large-scale evolutionary processes was vitally important to humanity, individually and collectively? Should I continue to pursue an academic career, given my intense interest in evolutionary ideas and my passion for 'big-picture' scientific thinking that is relevant to the human condition?

Against the background I have outlined, becoming a typical academic evolutionary scientist was not a viable option. It would have equated to choosing intellectual death. Instead, I chose life. This meant leaving academic science behind. The last place to pursue a career dedicated to the development of 'big picture' evolutionary science was at a university.

Instead, my plan was to find ways of earning a living that would give me free time to think and imagine. I would use this time to develop a general theory of evolution that covered not only biological evolution but also the evolution of all aspects of humanity, past, present, and future, including our psychology, consciousness, economic and political systems, ethics and values, and so on.

8.

Working for the Workers

When I left academia, I had two kinds of work in mind. Both were consistent with my overriding priority of developing a new theory of evolution. Given this objective, the work involved could not be physically tiring, and it had to give me the spare time to do heavy-duty thinking, reading, and writing.

But I had other goals. The work also had to provide a good middle-class income so that I could raise a family comfortably with a reasonable standard of living. I was not yet married, but I had a long-term girlfriend, and we were having a great time together.

The first possibility was teaching with its short daily working hours and long school holidays. The second was a government job in Canberra.

However, circumstances initially pushed me in a different direction. But it was not until many years later that I realized that this other direction was, in fact, critically important for driving my further cognitive development.

Of the two work options, my preferred choice was teaching. Now that I had a science degree, I could do a one-year university diploma, and emerge as a qualified teacher. Teaching jobs were plentiful, particularly in science and mathematics.

However, given my imminent marriage, I first needed to get a job for a year or so to earn enough money to tide me over while I completed the teaching diploma. In Australia at that time, university education was free, but I needed to build up some savings if my family was to have a decent standard of living.

The first job that caught my eye was as a trade union organiser with the Municipal Officers Association in Queensland. This union represented white-collar workers and professional staff in councils and other local government bodies. It seemed right up my alley.

As I have mentioned, it was fundamental to my nature to set myself against authority, particularly when power was being used to exploit people who were not in a position to defend themselves effectively. In my two years working for the federal government in Canberra, I had become the union representative for my fellow employees in the main department in which I worked. I enjoyed negotiating with the departmental managers and pressuring them to take employee interests more seriously. In particular, I designed and negotiated the introduction of one of the first flexible working hours schemes to be introduced in government employment at the national level.

I talked my way into the job with the Municipal Officers Union. I began organising members and dealing with disputes across Southern Queensland. I found it very fulfilling to advance the interests of employers by negotiating, arguing, manipulating, and pressuring the hierarchy in local councils.

Before long, I was also representing members before industrial tribunals. Australia has a system of arbitration tribunals that are empowered to act as an 'industrial umpire' in disputes between employees and their employers.

For example, when a dispute about wage increases cannot be resolved directly between employees and employers, either side can refer the issue to an industrial tribunal for determination by arbitration. The tribunal will hear arguments from both sides and impose its decision. It works like a normal court of law, but is less formal, and lawyers are mostly excluded from the proceedings to ensure that hearings do not get bogged down in technicalities.

If any dispute I was handling went to an industrial tribunal, I wanted to be the advocate for the members rather than having to rely on someone else coming in to take that role. The only way I could develop and control the strategies used in a dispute was to carry out all the key roles that were involved, including that of advocate in any arbitration proceedings. I did not want to have to convince others of the correctness of the often-complex strategies I would develop in particular disputes. I had already learned that most did not have the cognitive capacity to identify effective strategies.

The union was more than happy to have me develop the capacity to act as an industrial advocate. They saw that I was fully committed to developing the relevant skills and knowledge. They also could see that I was articulate, able to construct persuasive arguments, and could present a case confidently, with authority, and, as I was told, even with a touch of charisma.

The work was extremely challenging, and therefore demanded the development of greater skills and capacities. This was not limited to the need for further horizontal development. As I will discuss below, it also provided me with a powerful impetus for additional vertical development. It was also lots of fun, and very satisfying.

But at the end of two years, it was time to leave the union and start studying for my teaching diploma. It is usually nice to make another human being very happy, and I will never forget the smile of pure joy that split the face of the deputy head of the union when I told him that I was resigning my position. His all-consuming ambition was to take over as the head of the union. He was jubilant to see that the person who he believed was his greatest competitor was leaving. He obviously did not know that I had never shared his ambition at all, and that I was marching to the beat of an entirely different drummer.

I started the course for the teaching diploma at the University of Queensland in 1980. However, it did not take me long to realize that teaching was not the path to my goals. Due to deeply ingrained personality traits, I did not like being in a position of authority over anybody, including students. This was particularly the case given that my job as a teacher meant that I had to use this authority to uphold and impose values that I often did not agree with or follow in my own life.

For example, in my meandering life to that point, I had decided not to pursue many feasible career options that were highly valued in society. The world was at my feet, as I was told. But according to commonly-held values, I had squandered those opportunities. By the time I was doing the diploma, most of the kids who had been in my class at school were now doctors or lawyers.

This clash in values meant that I could not interact authentically and openly with students. But there was an additional problem. I found that teaching required long hours of concentrated work if it were to be done well. Competent and authentic teaching could not be accomplished within school hours. It required dedication and commitment. I saw that teaching would occupy my mind for long hours each day. It would not leave me the extensive time needed to develop a comprehensive new theory of evolution.

I pulled out of the diploma. I was now married with a family imminent. I needed to earn some money. So I got another job in the union movement.

This time, I worked for the biggest union in Queensland, the Australian Worker's Union (AWU). The AWU represented mainly semi-skilled and unskilled workers across most industries, from hospitals to sugar mills to mines to civil construction. Given its size, the AWU had considerable political power within the worker side of politics in Queensland and nationally. During the 1930s, half the leaders of the Queensland Labor Government had been members of the AWU.

I was employed as one of six industrial advocates. Our prime function was representing members in cases before arbitration tribunals. The work was extremely demanding, challenging, and highly pressured. The industrial advocate associated with a particular dispute inevitably developed the union's strategies and led the dispute from the union's side. This was because only the advocate had the legal and technical knowledge essential for developing effective strategies where the involvement of arbitration tribunals was possible.

The advocate would have to assess when strike action might be cost-effective, what form it should take, when and in what circumstances the union might seek to get arbitration tribunals involved, how to involve other unions that might be affected by the dispute, how to counter the various tactics used by employers, develop and lead media campaigns, and so on. All the way through a dispute, the advocate would organise and address mass meetings of the relevant union members, and sell them his strategies.

In large disputes in which strike action caused considerable disruption to the public, the advocate's strategies would also have to deal with inevitable interventions from the State Government. At the time, Queensland had a right-wing, anti-union State Government. It regularly intervened in any dispute involving large-scale strike action, threatening legal action and punitive measures. The Government found that a belligerent, anti-union stance was frequently to their political advantage.

The stakes were high, and the pressure immense. The imperative in all disputes was to develop cost-effective strategies and actions for members. However, full strike action is extremely expensive for the workers involved. They have to survive without pay for the duration of the strike. And there is no guarantee that their strike action will be sufficient to pressure the employers to make worthwhile concessions. This was particularly the case where the State Government intervened on the employers' side.

Given the complexity of large disputes, uncertainty bedevilled all strategizing. It was rarely possible to develop a strategy that was guaranteed to succeed. There were too many unknowns and too many different players and agendas involved. Strategies had to be continually adapted 'on the fly' as opponents acted in unforeseen ways and developed new tactics.

It was common for industrial disputes to end in disaster from the perspective of the members involved, even when significant concessions were won. Often, the costs of strike action far outweighed the benefits.

All this magnified the pressures and stresses on industrial advocates and others who led major industrial disputes.

Of course, many industrial advocates and other union leaders reacted to these challenges in the way that human beings normally do: drinking problems were rife, they avoided organising disputes even where members were agitating strongly for direct action, they looked for scapegoats and excuses if things turned bad, they developed a hatred of members that insulated themselves from complaints and criticisms, and so on, and so on.

But for me, it was precisely the kind of environment that I needed to propel my further cognitive development.

I was strongly motivated to succeed. This was central to my personality. From a very young age, I was driven to win every argument, every skirmish, every fight I got involved in. Super-competitiveness was central to my nature. My brother and sister remember that even as a five-year-old playing board games, I would do anything to beat them, including by bending the rules or by distracting them.

This went hand in hand with perfectionism and a powerful need to control my environment. It was not until many years later that I realized that this constellation of traits was a symptom of certain kinds of autistic predispositions. But whatever the causes, as a union advocate, I was strongly motivated to do my job perfectly. Avoidance was not an option. Failure was not an option.

Up until this point in my life, I had become adept at using what I have referred to somewhat metaphorically as First Enlightenment thinking. As I have outlined, this thinking is underpinned by the analytical/rational cognition that powered the rise of modern science. Its use is very effective for working out ways to achieve goals in circumstances that can be analysed and thought through. In other words, it works well when the circumstances that are being dealt with can be modelled and understood by relatively mechanistic thinking and mental models.

As I have also noted, the ability to use analytical/rational cognition to adapt effectively can be turbocharged by 'going meta', i.e., by using analytical/rational thinking itself to recursively improve its own ability to solve adaptive problems. In practice, this entailed my development of a capacity to see my analytical/rational problem-solving strategies as object. Once I could stand outside my thinking and see it from the outside, I could then assess where my thinking strategies were limited and how they might be improved.

My active development of this capacity had made my life very easy. I could easily work out how to achieve my goals, particularly where they involved challenges that could be analysed and thought through. With nearly all work in science, this capacity is all that is needed to function effectively. In relation to life, analytical/rational cognition is also very effective for achieving goals, provided you can avoid complex challenges that cannot be controlled with this level of thinking. However, avoiding complex social challenges is not much of a problem if, like me, you are self-centered and do not depend much on other people for your source of happiness.

Generally, I could use my turbo-charged problem-solving capacities to get whatever I wanted. I treated my pursuit of happiness as a series of problems to be solved.

My strategies for achieving happiness were not limited to trying to manipulate and control my external environment. I still remember an 'Aha' moment I had in my second year at university at a demonstration against the Vietnam War. I noticed a placard that attacked Richard Nixon. The placard contained a drawing of Nixon looking suitably tricky and insincere with the caption: "Would you buy a used car from this man?". Underneath was an image of a person responding: "Well, I bought a used car from him and I am happy." Underneath that was the response: "If you bought a used car from this man and you are happy, then you would be happy with your arse on fire."

In addition to the placard having the effect on me that was intended, it also evoked additional thoughts that were more impactful, at least for me. I saw that the placard could also be interpreted as identifying a particularly effective way of achieving happiness. If you could be 'happy with your arse on fire', your happiness would tend to be independent of external events and circumstances that might otherwise make you unhappy.

This stimulated me to ponder a number of issues: How could I develop the capacity to be 'happy with my arse on fire'? How could I remake myself so that happiness was no longer threatened by events that might be outside my control?

The possibility of intentionally developing such a capacity resonated somewhat with the approaches I had come across in my early teens in books about George Gurdjieff. As I have mentioned, Gurdjieff

73

advocated the need for individuals to work on themselves in order to change their psychological functioning. Unfortunately, the books I had access to did not contain explicit, understandable techniques and practices for achieving this.

But they were very clear that in order to develop higher capacities, individuals had to remake themselves comprehensively. This included developing the ability to change themselves to escape being jerked around by external circumstances, and to free themselves from the dictates of negative emotions and maladaptive conditioning.

In my late teens and early twenties, these ideas led me to develop several techniques that helped me be happy despite negative experiences and emotions. These techniques were very limited compared with the practices I learned later in life (as we shall see in detail, these practices involve radical acceptance and related methods for disengaging from negative emotions and prior conditioning). But the earlier approaches worked well for me at the time.

These techniques involved the use of cognitive reframing and related intellectual approaches. For example, I was able to take advantage of the fact that philosophy had failed to find ways to justify moral principles and norms. Philosophy was unable to construct rational arguments that could objectively justify what a person ought to do in particular circumstances.

Rationality had killed God, but was unable to replace religion's ability to ground human values and morals. Taking full advantage of this vacuum, I adopted an amoral stance that licensed me to do whatever I wanted, including flouting social conventions and norms. I went on to develop a comprehensive philosophical perspective that legitimated the self-centered pursuit of my interests without guilt or self-recrimination. For someone with my psychological predispositions, it was a very convenient philosophy.

This intellectual rationale meant that I was largely free to do what I liked. And I was very effective at manipulating my environment to achieve whatever it was that I wanted. Achieving happiness was simple and easy.

But this did not mean that I often acted in ways that hurt other people or disregarded their interests. A fundamental part of my nature at that time was to help others and to be authentic in relationships. My deepest psychological predispositions led me to be left-leaning politically. I was strongly motivated to oppose the abuse of power and authority, and to defend the exploited and the underdog. At this stage of my development, I had not yet set out intentionally to remake myself in ways that changed these fundamental predispositions.

What more did I need? I had it all worked out. My highly-developed, meta-analytical/rational cognition seemed to be able to take me wherever I wanted to go. Like most mainstream scientists, I tended to think at the time that I had the potential to understand all those parts of reality that were understandable. Also, like them, except for occasional intuitions and feelings, I was blissfully unaware that there was a higher level of cognition that would enable me to understand far more about complex aspects of reality, including human political, social, and economic systems.

At the time, I was even a tiny bit arrogant and obnoxious about my ideas and understandings. This particularly came to the fore when I was critiquing the arguments of religious people. Their belief systems were no match for the sword of reason that I wielded with abandon. I sometimes refer to this period as the 'Richard Dawkins' phase of my cognitive development. Fortunately, my development did not end at this point.

Working for unions and having to develop strategies that would work in highly complex, dynamic circumstances helped to blast me out of this complacency. This 'blasting' was essential. As I noted earlier, vertical development does not occur easily. It is extremely rare in current circumstances for an analytical/rational scientist to transition to Second Enlightenment thinking and metasystemic cognition during their adult life. In large part, this is because the levels that exist in cognitive development are dynamic equilibria. In systems theoretic terms, they are metastable. As such, they are resistant to change. They are constituted by a collection of capacities, skills, and beliefs that tend to be self-reinforcing and self-validating.

An individual at a particular level can see the limitations of levels below, but cannot see the advantages of the next higher level. As I have noted earlier, you must be at the higher level in order to see what its advantages are, and to see the limitations of the level below. This is because the higher level is able to model aspects of reality that cannot be represented adequately at the lower level. If a level of cognition cannot represent something, it cannot see it and learn from it. Consequently, it tends to be a slave to the lower level. You cannot see what is missing from your thinking. You cannot see what you cannot see.

For these reasons, individuals at a particular level will tend to dismiss critiques of their thinking that emanate from higher levels. Typically, they will reject attacks made from higher levels because they fail to meet the criteria used at their level to evaluate the validity of ideas and arguments. Often, they will be absolutely convinced that their position is correct, and that individuals at higher levels are delusional. This further entrenches them in their current level and inhibits vertical development.

Mainstream scientists who develop an interest in the emerging field of Complexity Science have not necessarily transitioned vertically from analytical/rational to metasystemic thinking. As I have mentioned earlier, a proper understanding of current complexity and systems science demonstrates that most of its findings are largely a product of analytical/rational thinking, not metasystemic. Its models are exactly what you would expect when a particular level of cognition is used to investigate phenomena that can be understood only by mental models at a higher level of cognition.

Nearly all modern complexity science represents an analytical/rational reduction of complex phenomena. It produces mechanistic models of aspects of complexity. When analytical/rational scientists attempt to investigate complex phenomena, they search for aspects of the phenomena that can be approximated by models that are analytically tractable and mechanistic. Such models will satisfy the relevant criteria used by analytical/rational science to assess whether theories and models represent 'true' science. When it produces such a model, analytical/rational science claims that it is developing Complexity Science.

In contrast, if metasystemic cognition produces models that represent complex phenomena adequately, often those models will not meet the 'truth criteria' of analytical/rational science. Such models will not be analysable or be able to be thought through. They will go considerably beyond the relevant data, identifying patterns and processes that cannot be represented or derived analytically. Furthermore, the models and the understandings that they produce will not be able to be described adequately in writing or diagrams. Metasystemic mental models of complex phenomena will tend to be rejected as unscientific by analytical/rational science.

A striking illustration of how individuals at these two different levels see each other's thinking was provided during an interaction between Bertrand Russell and Alfred North Whitehead in 1940. Whitehead and Russell had worked together at the beginning of the 20th century to produce the monumental Principia Mathematica. It was a heroic attempt to establish firm foundations for mathematics from an analytical/rational perspective. After this, they went their separate ways. Russell continued to use his brilliant analytical/rational thinking to address key philosophical issues. But Whitehead headed down the track of developing metasystemic cognition and used it to produce his process philosophy.

When they met on a podium in New York many years later, Whitehead summed up the outcome of their different cognitive trajectories. Whitehead said: "Bertie says that I am muddle-headed, but I say Bertie is simple-minded."

This comparison captures precisely how these different levels of cognition tend to see each other. Russell was a genius at the analytical/rational level, but failed to transition to the metasystemic level, except for occasional intuitions. As I have mentioned, he was my intellectual hero while I was reading in the Cairns Library, but as I developed cognitively, this did not last.

I will return to these issues in detail later in the book when I identify practices and methods that can be used to scaffold this difficult vertical transition from First Enlightenment thinking to Second Enlightenment thinking and metasystemic cognition.

My work as a union advocate provided the powerful motivations needed to drive my further vertical development. I was strongly motivated to succeed at my job, which demanded that I develop a capacity to strategize effectively in dynamic and complex circumstances. This strategizing could not be done effectively with analytical/rational thinking alone, no matter how successfully I had enhanced it recursively and consciously.

These factors combined to provide me with strong incentives to break free from slavery to analytical/rational thinking and develop metasystemic cognition.

If I had stayed in academia and pursued a career as a scientist, I probably would not have encountered such powerful incentives to develop my cognitive capacity. I would have remained in an environment where highly developed analytical/rational cognition would have been sufficient to produce a successful career. My scientific competitors would have been at the analytical/rational level. With that level of cognition, I could have achieved the rewards and accolades that come with a successful scientific career.

Of course, I would still have had the intuitions and predispositions that generated my passion for 'big picture' evolutionary theorizing and propelled me into conflict with the academic staff in the Zoology Department at the University of Queensland. I would still have been stimulated by the work of evolutionary thinkers like Teilhard de Chardin, and would have wanted to extend their insights and re-establish them on broader and sounder foundations. But I would not have been embedded in an environment where an inability to develop metasystemic cognition would have led me to fail to achieve my goals, repeatedly and painfully.

However, powerful incentives to develop metasystemic cognition do not create it overnight. This is evidenced by the fact that our current economic system generates many roles in which effective performance demands metasystemic cognition.

For example, the CEOs and senior executives of multinational corporations continually face complex challenges involving dynamically-changing competitive, regulatory, and customer environments. However, very few respond to these pressures by developing a strong capacity for metasystemic cognition. Cognitively, most are floundering and out of their depth. They are well and truly in over their heads. Fortunately for them, very few of their competitors can do better. However, once a reliable method for scaffolding metasystemic cognition becomes available, this will change rapidly. Possessing this capacity will rapidly become mandatory for work in the upper echelons of business, government, and academia.

When confronted with complex challenges, the first reaction of an individual at the analytical/rational level is often to double down on their use of analytical/rational thinking. Initially, this is what I tended to do in my union work. When facing the need to strategize in a complex workplace dispute, I would spend hours and hours in mental work, analysing and thinking-through different scenarios. I would envision possible strategies, possible responses, possible ways in which I could counter or take advantage of those responses, possible reactions to my counter moves, and so on, and so on, to the depth that was possible given my energy and time.

With simpler challenges, this could work very well. However, as the number of interacting participants increased, the growing complexity quickly overwhelmed my capacity to keep track of what was relevant. As complexity increases, the number of possible interactions between participants undergoes a combinatorial explosion. This rapidly overloads thinking strategies that require analysis of all relevant elements and that build mechanistic models. My response was to spend even more time attempting to model all this complexity mentally.

This was exhausting and stressful. However, with a lot of effort, it led to the discovery of at least some ways of handling complexity more effectively. In particular, I learned that word-based propositional reasoning is far too slow, ponderous, and simple to represent complex phenomena quickly and adequately.

I also discovered more about the kinds of mental processes that are capable of representing complexity. These processes rely far less on words or other symbols, and more on images and flashes of images. These images do not necessarily resemble the elements in the complex systems under consideration. Often, they are just fleeting glimpses that may sometimes be associated with a number of words, but in other instances, they may bear little relationship to words or what they point to.

Furthermore, I found that successful mental modelling of complex phenomena often demands that individual elements be represented collectively as patterns, structures, subsystems, and other larger-scale processes. These processes include the feedback loops and control structures identified by cybernetic approaches. This form of 'coarse-graining' can take away the need to represent each element individually, and reduces the complexity that has to be explicitly represented and kept track of in operating mental models. Appropriate 'chunking' and 'coarse-graining' can enable collections of individual elements to be represented by single mental images or flashes of images.

I got better and faster at developing strategies that were effective in complex circumstances. For example, I developed the capacity to assess very quickly how I might manipulate the participants in a labour dispute, despite their conflicting interests. I became adept at identifying what combination of things they could agree to that were also in the interests of the workers I represented.

I could walk into a room to participate in negotiations with the various parties involved in a dispute and almost instantly 'see' where the interests of each participant lay, identify where the various interests might overlap, and assess whether there was a sweet spot where the interests coincided with the interests of my members. In the absence of such a sweet spot, I could quickly assess whether I might be able to initiate strikes or other strategies that could change the interests of the participants in ways that would create a suitable coincidence of interests.

Increasingly, I could make these kinds of complex judgments very quickly, without having to think through the possibilities consciously and laboriously, step by step. I developed a general mental model of how to make these judgments and assessments. This general model could represent and simulate the structure and dynamics of a wide range of situations involving conflicts of interest. In any given dispute, all I had to do was to plug the specific details into my general model, and harvest its predictions and insights. This process became largely automatic, and much of it happened in a flash.

However, the process of learning and developing these kinds of techniques was long and slow. Nevertheless, I found it relatively easy to do better than others at strategizing in complex circumstances. This was sufficient for me to gain widespread recognition as a highly capable leader and strategist by others in my union, and in the industrial relations community more widely, despite my young age. But it was still very challenging to get it right in complex disputes. By the time I finished working for unions, I was far from being an accomplished metasystemic thinker.

Nevertheless, the difficulties I encountered when attempting to develop complex mental modelling had a very important implication for my central goal—developing a new theory of evolution. My struggles with these challenges generated the strong intuition that my thinking was too rigid, too mechanistic, and too structured to enable me to accomplish my goal. Increasingly, I realized I needed to disrupt and loosen up the thinking capacities that had previously served me well. From the perspective of First Enlightenment thinking, I needed to become more muddle-headed—but in a way that could make sense of complexity.

At the time, I did not know much about what I would need to do to achieve this. I knew that I had to become more intuitive and creative. I needed to be able to see large-scale patterns that cannot be perceived and understood with narrow analytical thinking alone. From my reading about the ideas of Gurdjieff and others, I had come to understand that to do so, I had to be able to disengage from incessant embeddedness in thinking and analysing, and instead let intuitions arise. However, I had no detailed understanding of how I could achieve such a capacity or what it might feel like to do so. Nevertheless, these intuitions increasingly primed me to be on the lookout for knowledge and practices that would enable me to achieve this.

81

A combination of factors led to the end of my work for the Australian Worker's Union and soon after, to the end of my career as a union advocate, organiser, and negotiator.

First and foremost, this was due to the realization that while I worked for unions, I would never have the time to develop a new, general theory of evolution. The union work was all-consuming, highly stressful, and very demanding. Advocates who took their work seriously burnt out quickly, and a disproportionate number died relatively young through stress-related disease, often exacerbated by alcohol abuse and smoking.

During my time at the AWU, I was smoking three packets of cigarettes a day. This took some doing. Often, I would use one cigarette to light the next. Frequently, I would wake up in the middle of the night to have a smoke. A friend from school days said that he and his family would amuse themselves watching the TV news by trying to spot me in reports of industrial disputes. He said that they could quickly tell whether I was involved in a dispute by looking for a plume of cigarette smoke rising from the group of negotiators, and then tracing it back to its origin.

Furthermore, I could see that my days with the AWU were numbered. I was a square peg in a round hole (Again!). The AWU was a conservative, right-wing union. In part, it owed its large size to the fact that employers preferred it. Faced with a choice, many employers would prefer to have their employees represented by a conservative union like the AWU than by a more militant, activist, left-wing union. Employers would often pressure their employees to join the AWU. In many instances, the AWU was criticized for being a 'bosses union'—a union that often served the employer's interests and sold out the interests of employees.

Through force of personality and intelligence, I had significantly changed the AWU's orientation, although, as circumstances were to prove, only temporarily. It was not that I had a wider, left-wing political agenda. It was just that I tried to win every dispute I led, and I was prepared to do whatever it took. No tactic or strategy was off-limits. For me, the end tended to justify the means.

Soon, the AWU was leading major industrial disputes in Queensland. This often led to conflict with the anti-union State Government, and many employers wondered what had happened to their tame union.

In my time as a union official, I played a major role in many serious disputes. I represented power station workers in the first-ever strike action that turned off electricity throughout most of Queensland. I organised a dispute that stopped garbage collection on the Gold Coast, Australia's prime national holiday destination. Then I went on national television warning tourists not to waste their vacation by holidaying there amidst rotting garbage. I led disputes that cut off the gas supply to parts of Queensland, stopped fuel supplies to much of Queensland (and went on television asking motorists not to make shortages worse by panic buying), cut water off strategically, closed hospitals (and used the media to request that any volunteers who took the place of employees not steal hospital cutlery and other equipment), and so on.

When I commenced my job with the AWU, I had little knowledge of its internal politics and its role as a tame-cat union. Up to that point, I was largely science-orientated and educated. I had little knowledge about the complex social and political processes that shaped how people and organisations behaved within society. Initially, I was so naïve that I assumed unions would always pursue the interests of their members and that union employees would do the same.

However, as I slowly began to understand the AWU's role in the State and how I was altering this radically, I also began to see that these changes were not sustainable. A militant AWU was beginning to lose the support of employers and governments. If it continued down this path, it could not continue to exist in its current form. It would become a much smaller union. This would reduce its political influence within the Labor Party (the political party in Australia that claims to represent worker interests). This was because the AWU's power within the Party depended on the number of votes it was allotted, and these were allocated based on its size.

Increasingly, I began to realize that it was in the interests of many powerful forces outside the union to do what was necessary to

83

return the AWU to its previous role. I saw that the simplest way for these forces to achieve this was to get rid of me from the union. I was the single cause of this threat to their interests. Furthermore, doing so would also be in the interests of many who held power within the union itself, who benefitted from the previous status quo, and who would lose out if the union continued down this path.

I remember a particular event that drove home these realizations. I was talking to a journalist for a major Queensland newspaper on the phone. I had given him a great story about a hospital dispute the day before. Something prompted me to ask him how he saw the AWU and the way it went about representing its members. He said that he saw the AWU as a powerful, militant, left-leaning union that was prepared to fight strongly for its member's interests. When I heard this, I realized that it meant that my time at the AWU was strictly limited. Soon, they would come to get me.

The final factor in ending my union employment was that I was also becoming a bit uncomfortable with what I was doing as a union leader. I was effective at winning most industrial disputes in which I was involved. I was able to develop complex and innovative strategies designed to manipulate opponents' interests and stay steps ahead of their counter-moves. I was good at maximizing the powerful forces that I could bring to bear on employers to persuade them to give in to our demands. But I was dealing with symptoms, not causes.

I was adept at rectifying specific instances of the exploitation and underpayment of members. But I had no strategy for dealing with the larger-scale forces in society that produced these particular instances of exploitation and many more. No matter how successful I was at rectifying specific instances, these societal forces would continue to generate exploitation. At the time, I lacked the mental models that would enable me to see how society could be rearranged in such a way that exploitation would no longer emerge.

It hit me that I was actually operating like the Red Brigade 'knee cappers' who were terrorizing rich families in Italy at the time. A key tactic of theirs was to machine gun the knees of the children of the rich and powerful. Like me, their ends justified their means. Also, like me, they had no feasible, 'big picture' plan for reorganising society appropriately. From the perspective of the society in which I was embedded, but to a somewhat lesser degree, I was also out of control and dangerous, just like the 'knee cappers'.

This combination of factors caused me to start considering employment options outside the union movement. In particular, I looked for work that would give me a good middle-class income and enable me to provide for my family (I now had a wife and two young daughters). However, it also had to leave me with the time and energy to pursue my evolutionary theorizing.

Broadly, it seemed that returning to government employment in Canberra was the best option for meeting both these criteria.

But as things turned out, I did not leave the AWU at a time of my choosing, with another job already lined up. Instead, I got involved in an internal dispute over how a particular staff member was treated. Half-heartedly, I organised a strike of union officials to protest against this treatment. Within a day, the AWU executive met and sacked me.

Apparently, they thought that I was attempting to take over the union. They had no idea that I was looking to leave. The day after I was fired, I bumped into the head of the union in the street. He told me melodramatically (and rather narcissistically), that a part of him had died when I gave him no alternative but to sack me. But he was like that. And we had shared some extraordinary experiences together on the battlefields of workplace disputes.

Within a few days, I commenced a six-month contract to work for the Queensland Nurses Union. This helped tide me over until I got a government job in Canberra. Although I did some valuable work for the nurses, a significant part of the reason for their support was compassion. They assumed that my dismissal was political and unjustified. This was because I had got to know some key staff in the Nurses Union previously and they knew my capabilities. Knowing that being sacked could be a very unpleasant and challenging experience, they decided to help me with a new job.

HUMAN SUPERINTELLIGENCE

My time with the Nurses Union was the only period of work in my life in which I felt genuinely cared for. It was not something I looked for in employment, but it was a very pleasant experience.

9.

Working for the Government

In early 1985, my wife, two young children, and I headed over 1000 km south from Brisbane and Queensland to live in Canberra. As I have mentioned previously, it is Australia's national capital, the seat of the federal Government, and largely a city of government employees.

My new job was as a Class 9 officer in the Department of Industrial Relations. My work entailed providing the Government with policy analysis and advice about how Australia's industrial relations system should be regulated. This included assisting with the drafting of any legislative changes that the Government decided to pursue.

As I have indicated, the goal of my move to Canberra was not to have a successful public service career. Rather, my overriding objective was to develop a new theory of evolution.

Several times during my government employment, I attended meetings that began with an invitation to participants to say a little about their backgrounds and goals. I often told the meeting that my central purpose in working for government was to develop a new, 'big picture' theory of evolution. The other participants in the meeting invariably thought I was joking.

My salary at the Class 9 level was enough to fund a good, middle-class lifestyle for my family. While our daughters were young, we decided that my wife would look after them at home, rather than engage in paid work in her profession as a registered nurse. We had found previously that when we both worked full-time in paid employment, we had less time with each other than we liked, and life was extremely busy. Now that we had two young daughters, I was more than happy that she was willing to be a full-time mother.

HUMAN SUPERINTELLIGENCE

After a couple of years, we bought a block of Canberra's cheap land, and had a comfortable family home built in an established area with good public schools that were in easy walking distance. One of the great advantages of living in Canberra was its excellent public facilities, including schools and hospitals. Many of the public schools were at least as good as expensive private schools in the main capital cities. Of course, the quality of the public amenities in Canberra often produced envy in visitors from other capitals. I used to make it a point when I met other Australians to thank them profusely for the taxes they paid to help fund our facilities.

Early after moving to Canberra, I experienced a series of challenges that drove me to look again at some of the techniques advocated by Gurdjieff and others for 'working on oneself'. Much of Gurdjieff's work on oneself was directed at becoming what he referred to as a 'self-evolving being'. Broadly, as I mentioned earlier, this described individuals who could change themselves at will, freeing themselves from past conditioning, negative emotions, and other predispositions that limited their effectiveness in the world.

Deep at the back of my mind, I had carried the idea that a self-evolving being would be able to achieve the ultimate: they would be able to be happy with their arse on fire. No matter how challenging the external circumstances, happiness could be achieved by working on oneself internally.

As I have indicated, I had been intrigued with these ideas since I was a teenager. Previously, however, I had little need to attempt to put them into practice because life had been relatively easy for me. Union work had been very stressful at times, but I could generally strategize my way out of difficult circumstances fairly quickly. Furthermore, although I did not recognize this then, my incessant thinking and strategizing helped suppress unpleasant emotions. Embeddedness in thinking tends to fill the limited bandwidth of consciousness and reduces awareness of emotions.

For these reasons, union challenges tended to drive me to develop intellectually, rather than to work on my emotions.

But this changed when I decided to give up my three-packet-a-day smoking habit. Recognising its detrimental impact on my health, I had tried to quit smoking twice already in my final year working for the AWU. However, given the stress I was under, I soon went back to smoking.

But when the AWU fired me, I resolved to ensure that something good would come from this experience. I decided never to smoke again. The motivation I got from this was powerful, and I have not smoked since, except occasionally when dreaming. But it was a rough ride for many months, and I looked for techniques to make it easier.

Furthermore, I had always loved sugary treats, and took to eating two family packets of lollies each night as a substitute for smoking. Soon, I was putting on a lot of weight. The sedentary and largely stress-free work of policy development did not help either.

I realized that I also needed to overcome my sugar addiction when I was running to my car to avoid a rain shower. The rain had just started when I was about ten meters from my car, but this short sprint left me gasping for breath and having to rest. Clearly, I had to make another significant change to my behaviour.

Unfortunately, neither Gurdjieff nor his followers who wrote about their experience with him were very clear about the nature of the practices and techniques that could train the capacity to be self-evolving. Furthermore, it seemed that hardly any of his followers (if any at all) had actually implemented his techniques and become self-evolving beings.

However, after reading Gurdjieff's books and other resources, and also after experimenting on myself, I came to the conclusion that a process of silent 'self-observation' was at the core of his practices. The general idea was to engage in self-observation in real-time, while one was experiencing cravings for cigarettes, or food, or a negative emotion.

The practice involved giving 'bare' attention to the actual sensations produced in the body by cravings or emotions. This attention was 'bare' and 'silent' in the sense that while self-observers were passively witnessing their sensations, they would not think about the experience. Only awareness was involved.

The effects of this silent self-observation could be enhanced by giving attention to the details of the sensations, e.g., their precise position in the body, their intensity, any changes occurring in them, and so on. The practice involved just sitting with and accepting the sensations—i.e., not reacting to them in any way, in action or in thought.

I found that this kind of practice was indeed capable of taking away the unpleasantness of cravings and negative emotions. I could stay relaxed as I watched the cravings arise and then dissipate. They became just harmless sensations, coming and going. I did not have to eat or smoke to get rid of the unpleasantness of cravings. I could resist temptation, with ease.

I also found that this approach worked even for the most intense pain I had experienced up until then—at the dentist. If I relaxed and experienced the pain as mere sensation, it was not painful any longer, regardless of whether the dentist had given sufficient time for the local anaesthetic to take effect.

I used to look forward to going to the dentist as an opportunity to practice this technique. For further practice, I would tease myself with food cravings. I would open the fridge's freezer compartment and look in at the ice cream, coming into the present and watching the cravings closely as they strengthened. I would stand there until the cravings were just sensations in my body that no longer motivated any action. Then, I would close the fridge and walk away.

There seemed to be two mechanisms at work here—one operating 'in the present' as I have described, and the other in the longer term. This second mechanism seemed to be able to rewire or recondition myself permanently.

The second technique made sense in the context of the elementary learning theory I first encountered in the book 'Psycho-Cybernetics' and then in Psychology 101 at university. By accepting negative feelings and sitting with them, responses to those feelings that had been learned previously but that were no longer helpful would be extinguished. The learned connection between the feelings and the reaction would be broken. The feelings would no longer evoke the unhelpful responses.

For example, negative emotions can trigger eating behaviours that tend to remove the unpleasant feelings produced by the emotion. Eating is used to self-soothe. However, instead of acting out this conditioning, individuals can simply give bare attention to the cravings and accept them. If this is done repeatedly, eventually the conditioned response will dissipate and lose force, permanently.

In later years, I learned that clinical psychology already uses similar applications of learning theory in diverse forms as a tool for changing unhelpful learned behaviours. An example is exposure therapy, which can extinguish learned fears that manifest as phobias (e.g., fear of flying in airplanes). I realized that the addition of silent 'self-observation' to these diverse approaches could turbo-charge their effectiveness.

These techniques worked effectively for stopping smoking and for regulating eating. But once I overcame the challenges that motivated their use, I moved on to other things. It was years before I experimented further and worked out in depth how self-observation and related approaches worked. Once I understood this, I was able to see how these techniques could be further enhanced and used more widely. Eventually, I grasped how they could be used more generally to enable individuals to free themselves from the dictates of their evolutionary and social past, as well as their upbringing. I learned how to scaffold the capacity to become a self-evolving being.

Soon after I moved to my new job in Canberra, I began working earnestly on developing a new 'big picture' theory of evolution. I used several strategies. First and foremost, I had to get across the work of others that was relevant to my project. This meant reading everything interesting I could find about large-scale processes and patterns in evolution. This required focusing not only on descriptions of these patterns, but also on attempts to identify the systems-level causal processes that produced them.

I developed two other strategies that were designed to ensure that this reading was as productive as possible for my goals. The first was to hold in my mind as my reading proceeded several key unresolved evolutionary problems. I had accumulated these at university and during previous reading. They were issues that evolutionary science fell far short of solving adequately at that time. These were the irritations I hoped would eventually produce pearls of wisdom for my new, general theory of evolution.

The most significant of these (and ultimately the most productive) was to explain how cooperation and altruism could evolve, given that self-interest was generally expected to prevail in evolutionary competition. Richard Dawkins had outlined this issue very clearly in his popular science book, 'The Selfish Gene'. He argued convincingly that organisms that use some of their resources to help others without any benefit to themselves, will tend to be eliminated by natural selection.

This was a very significant issue for previous attempts to develop 'big picture' theories of evolution. Most notably, Teilhard de Chardin claimed to have identified a large-scale trend in the evolution of both matter and life. He suggested that this trend led to the progressive integration of matter and life into larger-scale organisations. This occurred through the repetition of a step-wise process in which smaller-scale entities were integrated into larger-scale ones. For example, elementary particles were integrated into atoms, atoms into molecules, molecules into the first simple cells, simple cells into the more complex eukaryote cells, complex cells into multicellular organisms, multicellular organisms into animal and human societies, and so on.

Teilhard's failure to identify plausible causal mechanisms that would produce this pattern resulted in modern evolutionary science not taking his ideas seriously.

Another key challenge that I held in my mind as I read was whether a unified theory could be developed that could explain both biological and human evolution as special cases of a more general and abstract theory. Such a theory might, for example, demonstrate that the mechanisms that produce evolutionary change have themselves evolved. Within this broader framework, gene-based evolution and cultural evolution might be seen as specific examples of wider processes.

A further potential for such a unified theory would be to produce a comprehensive understanding of human evolution that can account for the evolution of all aspects of humanity. This would include the evolution of our political, social, and economic systems, as well as human psychological evolution and the evolution of intelligence.

Ultimately, a comprehensive theory of evolution seemed to me to have the potential to provide an overarching framework for all other sciences and for what is now called the humanities. It was also likely to enable humanity to understand how we needed to evolve in the future in order to continue to survive and thrive.

An additional major challenge was to understand within an evolutionary framework what is commonly referred to as spiritual development and attaining higher levels of consciousness, including enlightenment.

If this was not enough, I also wanted to pursue the intuition that a comprehensive understanding of how past evolution has shaped humanity and will also shape us in the future could provide us with an objective method to found our values, morals, and goals. In particular, it could enable us to objectively answer the key existential question that faces all of us: How should we live our lives? What should we do?

My third strategy had two components, with one meta to the other. First, the primary purpose of my reading was to motivate me to build my own mental models of large-scale evolutionary processes and patterns. I did not read just to understand the models developed by others. Instead, in relation to every issue of interest, I focused on developing my own models. This was aided by the key evolutionary challenges that I held in mind as I read and thought about my reading. In this way, I used my reading to generate novel thinking, not just to recombine the ideas of others.

The meta-component involved using my model-building skills to enhance my capacity to build effective mental models of complex phenomena. My goals were not just to build novel mental models that addressed key challenges in evolutionary theory—they were also to improve my capacity to build such models.

For example, when I found a thinker who was addressing issues similar to those that engaged me, I would put the book down and build my own mental models of the relevant issues from first principles. As far as possible, I started this model building from scratch, thereby avoiding

93

being biased by existing knowledge and assumptions. Then, I would pick up the book again and compare my models with those developed by the book. If its their models were superior in any respect, I would then consider how I could have used a better approach to build my models. The result was that I enhanced not just my models, but also my meta-models. I did this repeatedly whenever I came across new thinkers who had developed a different approach to the relevant issues. I recursively improved both my models and meta-models.

The goal of my reading was not just to understand the existing state of evolutionary science and to take current knowledge as a set of foundations to build upon. Consequently, I was not afraid to explore ideas outside of accepted science to stimulate my modelling and meta-modelling. In fact, intelligent mystical and new-age ideas often proved far more fertile for stimulating new models than mainstream science did. However, the resultant models often bore little resemblance to the ideas that stimulated them, particularly once they were cleansed of woolly thinking and mysticism. Nevertheless, I could not have developed these models if my thinking had remained firmly rooted in current science.

As I progressed, it became even clearer to me that existing science could not deal effectively with complex phenomena, such as those that manifest in large-scale evolutionary processes. The criteria that mainstream science used to assess whether particular approaches were sufficiently rigorous to be accepted as science, could not help but be violated by genuine systems thinking.

For example, complex systems models almost always go far beyond the relevant data. Furthermore, reductionist approaches cannot work for truly complex phenomena. As the number of interacting components increases, analytic models are quickly overwhelmed by combinatorial explosion and analytical intractability. I soon discovered that I had to look on the fringes of science and far beyond.

My explorations took me to the work of thinkers such as Gregory Bateson, Stafford Beer, cyberneticists more generally, key thinkers in general systems theory, Tolstoy, Ken Wilber, Madam Blavatsky and other theosophical writers, Gurdjieff, Steiner, Heidegger, Nietzsche, and many others. And my reading led me to discover that there were thriving communities of systems thinkers outside the 'hard' scientific disciplines in human organisational theory, economics, and even architecture.

These strategies also contributed significantly to implementing my intuition that to develop a new 'big picture' theory of evolution, I needed to de-structure and de-mechanise my thinking. This intuition told me that my thinking was too analytical, too reductionist, and too linear. Although this kind of mechanistic thinking had proven very effective for me in many areas in the past, it seemed that it would be far less creative for generating complex models of evolutionary processes.

At that time, I had no specific strategies for breaking up my analytical thinking and preventing it from crowding out the kind of intuitive, 'big picture' thinking I needed for my grand theorizing. It was not for another five or so years that I began to realize that there were specific practices that could be adapted for this purpose. These practices could enhance my ability to integrate intuitions and pattern-recognition capacities into my thinking.

However, at this time, I was aware that certain actions I could undertake in my daily life could facilitate this in specific cases. For example, sleeping on problems, walking in nature, having a shower, leaving issues alone for a few days, taking breaks, and even a bout of flu could lead to novel insights. Eventually, I came to understand the nature of the psychological processes that underlay these approaches and that facilitated the insights they produced. This enabled me to adapt and develop specific practices that could maximize their effectiveness. In time, I learned how to turn them into cognitive technologies.

Eventually, I realized that the reading strategy I had adopted was, in fact, the only way in which the writing and ideas of genuine systems thinkers could be understood in their full complexity. It is only by building, testing, and improving one's own models that one can understand the complex models that systems thinkers used to generate the ideas that they have written down. You will not be able to understand their models simply by reading and understanding their writing in a straightforward manner.

This is because of the limitations of written language. Writing is intrinsically sequential, mechanistic, and linear. It cannot capture dynamical complexity. It can effectively describe analytical/rational thinking, but not genuine complex thinking. Words do not even exist for many of the processes that constitute complex reality.

Consequently, when writing a book or an article, the best that a systems thinker can do is to write about aspects of their models that are capable of being represented using written language. They are unable to write down adequate descriptions of the complex models themselves. Instead, when systems thinkers write about their complex ideas, they have no option but to reduce them to compilations of analytical/rational argument, mechanistic models, and supporting evidence. Their complex, dynamic, mental models cannot be described adequately in writing.

Consequently, if a reader is to understand the complex models that a genuine systems thinker used to generate their book or article, the reader has to reverse this process. The reader needs to reconstruct from the written, analytical/rational account the complex models that the thinker used to generate the ideas. Of course, they will not be able to do this unless they have achieved or are developing the capacity to be a genuine systems thinker themselves.

The demands of my day job helped me in important ways to succeed with my evolutionary project. In particular, it encouraged me to use my growing ability for recursive self-improvement to enhance my writing skills. As I have mentioned, English had always been my worst subject at school, by far. I could not get 100 percent as I could in maths and science. This is because the tasks required in English could not be reduced to a set of rules that I could learn and apply faultlessly. Rules existed for some areas of English, such as grammar and spelling, but there were many exceptions to the rules. Identifying clear rules that could assist in writing essays was even more challenging. At school, I made very little progress in developing algorithms that could enable me to construct 'well-written and interesting' essays.

However, my government policy work demanded that I be able to write about complex ideas in a simple, concise, and easy-to-understand fashion. The central task to be performed by a policy advisor was to produce briefs on issues where the government might need to establish a new position or defend an existing position. In most cases, the briefs were written for the Minister responsible for the Department. The briefs had to be clear, short, and easily digested by a Minister who was always extremely busy, had little time to read, but who had to be across a diversity of complex issues so that he or she could respond coherently when questioned by an often-antagonistic media.

The most frequent form of briefing required by the Minister involved the preparation of Possible Parliamentary Question Time briefs. Whenever Parliament was sitting, the Opposition had an opportunity each day to question Ministers and the Prime Minister about their policies. Typically, questions were intended to expose flaws in the Government's approach. The briefs provided to the Minister to prepare him for possible questions often had to be provided at very short notice. Issues might have been raised in the morning newspapers, and it would be necessary for the Minister to have his briefs about them by 11 am.

When training new staff, I often drew on my background in Zoology to use a penguin metaphor to help them understand their role in preparing briefs for the Minister. The mother penguin would leave the young in the nest and travel widely to collect various food items. She would masticate these thoroughly to produce an almost pre-digested slurry. When the mother returned to the nest, she would vomit this slurry into the open mouths of her young. I would tell new staff that if they wanted a successful career in policy development, they should see themselves as a mother penguin, and the Minister as one of their young.

In response to these work demands and their pressures, I quickly learned how to structure complex written material, utilize formats that worked best for particular purposes, make material easily digestible, know what phrases and substructures were useful in particular circumstances, and so on. I recursively improved these skills at every opportunity. Most days during the lunch break, I would walk for an hour for exercise. In the early days of my new job, I often wrote and rewrote a brief in my head. Whenever I came up with something effective, I would identify the general principle that underlay it so I could use it in many other situations.

Work got progressively easier. I could get across the relevant subject matter quickly. Once I had identified and internalised the relevant writing skills, I could deliver what was wanted in a short time. This left me more time to spend on my evolutionary work.

However, as you may have noticed, my work did not demand that I develop the ability to write creatively and evocatively. I never did work on myself to develop those skills. You will have to settle for clarity.

Another benefit that my employment provided for my evolutionary theorizing was direct experience of the operation (and malfunctioning) of the complex systems that manifest as bureaucratic organisation. I have mentioned that my reading program soon took me through the readily-available papers and books that applied complex systems thinking to evolution. Consequently, I had searched far and wide for disciplines that were less allergic to intelligent and creative thinking about complex phenomena. I found that human organisational science was such a field. I discovered that it had long valued general systems and cybernetic approaches to understanding human organisation, including businesses and bureaucracies.

My employment provided me with direct experience in human organisations that I could use to test my relevant mental models as I developed and improved them. For example, I soon realized that the Division of about 60 people I worked in contained many highly intelligent and educated people. Most held legal qualifications because much of the work involved interpreting existing legislation and drafting new provisions. However, my experiences at work soon raised an issue that often features in complexity science: it is frequently the case that the whole is far more than the sum of its parts. An organisation often has emergent capacities that enable it to achieve outcomes that cannot be produced by the parts acting independently.

What struck me and what demanded explanation was that this was not the case for the Division in which I was embedded. Very often, I found that the organisation as a whole was capable of far less than the sum of the abilities of its staff. This organisation of highly-educated and talented employees tended to produce collective outputs that were often less competent and insightful than the sum of the capacities of the individuals involved.

This and related curiosities drew me into developing mental models of how my organisation tended to self-organise incompetence rather than quality work. I wanted to understand how so many smart and capable people could produce so little of value. How was it that they could produce outcomes that were not actually wanted by any individual within the organisation?

These issues led me to consider the nature of the incentives and disincentives that the organisation provided to its members. I began to see, for example, that these incentives tended to reward risk averseness; that this is particularly the case in bureaucracies in which the basis for all decisions is recorded and available for re-evaluation and attribution of blame for years into the future; that cooperation and collaboration often were not in the interests of employees even though they could potentially produce great benefits for the goals of the organisation (e.g., if employees kept specialist knowledge and skills to themselves, rather than sharing them with others, they could become indispensable to the organisation); that the levels of incompetence at higher levels were much greater than predicted by the 'Peter Principle' (this Principle predicts that individuals will continue to be promoted to higher levels until they get to a level at which they are clearly incompetent. However, the Principle fails to recognise that in a mature organisation, this will not stop the promotion of individuals to even higher levels of incompetence. This is because the continued operation of the Principle eventually means that those who decide on promotions are already at their level of incompetence and consequently will fail to recognise the incompetence of those applying for promotion); and so on.

These and subsequent experiences propelled me toward developing a general theory of organisation. In particular, it identified how organisations of all kinds evolve and self-organise. The theory was general in the sense that it applied to organisations of entities at all levels: e.g., the organisation of molecular processes into the first cells, of cells into multicellular organisms, of organisms into animal societies, and of humans into tribes, corporations, nation-states, and so on.

After about a year of ostensibly working on legislative policy issues, I was successful in getting a promotion to a Class 10 position in another Division within the Department of Industrial Relations. The attraction of this job was that I got my own office. This enabled me to position my desk so that no one could look over my shoulder and see what I was reading or writing. I also had additional staff to whom I could delegate work.

The work involved attempting to ensure that semi-government bodies did not embarrass the Government by breaking the Government's guidelines that restricted what pay raises the bodies could give to their employees. I was responsible for regulating several of these bodies, including Australia's public broadcaster (the ABC), the bank that was originally established by the national Government (the Commonwealth Bank), and the Australian National University.

The Senior Executives who ran the Division were never very clear about the details of what the other Section Heads and I were to do to achieve this goal. They never provided us with any powers that would enable us to align the interests of the semi-government bodies with those of the Government in this area. Without some source of power, we could not even find out easily what the bodies were doing about wages and conditions for their employees. Whenever they found it within their interests to breach the Government guidelines, they were incentivized to hide their breaches from us.

When asked by friends and others what my work involved in practice, I used to say that I had only worked in the Division for six months or so, and had not yet been there long enough to be told, or to work it out for myself. I was joking, but not by much.

Given my evolutionary goals, it seemed like the dream job. However, I succumbed to temptation after I had been there for less than a year. I came across a newspaper advertisement for a position that really did seem like my dream job, given all my interests. It was for the Manager of Australia's North Eastern Fisheries, based in the Australian Fisheries Service in Canberra.

At first, it seemed ridiculous to consider that I would have a chance of getting such a job. I had never worked in fisheries management, and had done only one unit about fisheries in my university degree. I dismissed the thought of applying. But over the next few days, I began to see that I could make a case for the job that might get me considered seriously.

For example, the job was at the Class 11 level, only one above the level I had already attained. Furthermore, because the position was based in Canberra, an inland city a couple of hours' drive from the nearest coastline, it was not likely to be an attractive location for marine scientists. Part of my reason for moving to Canberra had been to live somewhere far away from fishing opportunities that could distract me from my evolutionary theorizing. This factor could work strongly in my favour in getting a job as a fisheries manager based in Canberra.

So, I applied for the job and went to the interview. I made the case to the selection panel that although I had never actually worked in fisheries management, the work I had done previously contained all the key skills and capacities that were required of a fisheries manager. I outlined in detail how my employment record demonstrated that I could perform these capacities at a very high level.

In preparation for the interview, I had also spent a couple of days studying the economics and biology of fisheries management, so I was able to give them impressive answers to the questions they asked about fisheries management. And, of course, I had actually been a licensed Master Fisherman operating in some of the relevant fisheries.

HUMAN SUPERINTELLIGENCE

10.

Back to Fishing

A month later, I was the Manager of Australia's North Eastern Fisheries. I was responsible for managing the Commonwealth fisheries from Sydney to and including the Torres Strait. This comprised all fisheries in the Torres Strait and the commercial fisheries between the Strait and Sydney that lie outside Australia's continental shelf. The relevant State governments managed the fisheries on the shelf.

Of these, the Torres Strait fisheries were the most interesting and challenging to manage. Because the Torres Strait and its Islands were within the territorial limits of both Australia and Papua New Guinea, an international treaty had been agreed by the two countries to resolve the overlapping claims. The Treaty also took account of the fact that the Indigenous people of the Torres Strait and of the adjacent areas of PNG (who differed somewhat in racial origins), had both historically intermingled in the Strait, and used the fisheries and other resources. Furthermore, Torres Strait Islanders are all Australian citizens, even though some of their islands are only a few kilometers from PNG.

In relation to fisheries management, the Treaty prioritized traditional fishing by the indigenous peoples over commercial fishing.

This all meant that parts of the fisheries in the Torres Strait were managed solely by Australia, parts solely by PNG, and parts jointly by Australia and PNG.

The relevant Commonwealth and State Fisheries Ministers exercised Australia's responsibilities jointly. The Ministers were advised by a Consultative Committee that included representatives of Torres Strait Islanders, commercial fishermen, fisheries scientists, and both the State and Commonwealth fisheries managers.

As the Commonwealth Manager, I was ultimately responsible for advising the relevant national and State Ministers about how the fisheries should be managed, in light of input from the Consultative Committee. In relation to the fisheries jointly managed with PNG, we would travel to PNG a couple of times a year to resolve relevant management issues.

Even without these complexities and potential conflicts of interest, fisheries management is notoriously difficult. Most commercial fisheries around the world are heavily overfished, and many have collapsed. Almost none have been managed to maximize economic returns in a way that is sustainable in the long term.

The challenges inherent in fisheries management result from what is known as 'the tragedy of the commons'. This dynamic arises when there is unrestricted access to a natural resource such as a fishery. In such a case, it is in the immediate interests of each fisherman to take as many fish as he can catch profitably. Individual fishermen might realize that if they all continue to fish on this basis, the fishery will likely collapse, and all fishermen will lose. However, if any individual fisherman attempts to prevent this by voluntarily limiting what he catches, he will lose personally, and his restraint will not protect the fishery. It will remain in every other fisherman's immediate interest to continue maximizing their catch. They will realize that if they were to restrain their catch voluntarily, it would reduce their income but achieve nothing. This is the tragedy.

In general, this problem can be solved only by setting up a coordination mechanism that will require all fishermen to adhere to appropriate limitations to their catch rates. Significantly, since it will also be in the immediate financial interests of all fishermen to break the restrictions if they can get away with doing so, the coordination mechanism must also include an enforcement process that punishes any who breach the restrictions. If an effective mechanism is implemented, it will tend to align the immediate interests of all fishermen with the longer-term interests of the fishery as a whole. The tragedy of the commons will have been overcome.

The biggest challenge in fisheries management is to get the right management mechanisms in place. This is difficult in Australian fisheries management because the processes that are used to decide management arrangements are generally biased in many ways. They do not guarantee that effective management is implemented.

Broadly, this is because the interests of those involved in establishing management rules are not necessarily aligned with those of the fishery as a whole. Often, their interests conflict with successful management. Furthermore, the particular interests that end up being served by management arrangements tend to be those that are the most powerful.

For all significant Australian fisheries, the Minister is the ultimate decision-maker. The Minister is typically advised by the Fishery Manager and a Consultative Committee that comprises representatives of relevant interest groups, including representatives of the commercial fishermen. However, the problem is that the representatives of commercial fishermen tend to support their own individual interests, not those of the fishermen they are supposed to represent. They typically support management arrangements that are biased in favour of their own particular circumstances—e.g., the size of their boat, their fishing method, the port they fish from, and so on.

Fisheries scientists are typically biased toward supporting the kind of research that they specialise in. This will ensure that they and the institution they work for will likely end up with the research funding. The Minister is often biased in favour of taking whatever decisions are in the political interests of himself and his political party. The fisheries manager will tend to make decisions that are best for his career in the short term. He hopes he will have moved on to another position by the time the fishery collapses. And so on.

If a fishery were to be managed effectively, the Manager had to ensure that management arrangements were put in place that were often contrary to the interests of powerful players in the fishery. I used to describe the job of a fisheries manager as standing between a herd of starving pigs and their trough, trying to prevent them from eating everything at once.

It was not until years later that I came across the work of Elinor Ostrom. She studied the kinds of management arrangements that communities exploiting a natural resource would tend to put in place to overcome tragedies of the commons. Broadly, her research suggested that management restrictions that enabled the resource to be utilized fairly and sustainably tended to self-organise through consensus. She was eventually awarded a Nobel Prize for her work on these issues.

Ostrom's findings proved very popular amongst those with post-modernist sensibilities. They tend to abhor forms of organisation that rely on power and hierarchies, and hope that problems like the tragedy of the commons can be resolved by negotiation amongst equals. They readily accepted Ostrom's hypothesis that the members of the relevant communities possessed the foresight to see that appropriate restrictions would need to be applied to users of the resource in order to ensure its sustainability in the long term.

My direct experience was at odds with these hopes and dreams. In my experience, exploiters of resources (including Indigenous ones from both the Torres Strait and PNG) would use whatever power they could muster to advance their own particular interests at the expense of others. Ostrom's predictions might apply where all the users of a natural resource were equal in power both in the short and long term. However, it is difficult to imagine how this condition could be met in the real world, particularly over long periods of time.

Furthermore, Ostrom's research tended to disguise the necessary involvement of top-down power in enforcing whatever rules are made to manage a natural resource. In general, exploiters will not abide by management rules unless they are sufficiently punished if they flaunt those rules. This punishment will not be effective unless it is implemented by a process with sufficient power to punish all those who break the rules, without exception.

In short, fisheries management requires a system of powerful governance and enforcement if it is to be successful. Once Ostrom's research is properly understood and her downplaying of the role of power is seen, it is clear that her findings reflect these necessities. Her research does not conflict with recognition of the central role of power in the general models of the evolution of cooperation that I was to develop in coming years, and that were significantly influenced by my fisheries experience.

The challenges that I faced as a fisheries manager were similar to challenges that the evolutionary process has faced many times during the evolution of life on Earth. How do you get self-interested individuals to cooperate to act in the interests of a larger-scale organisation when it is against their immediate interests to do so?

When life first emerged as simple proto-cells, each living entity was less than a millionth of a meter in diameter. But as I mentioned earlier, living entities did not remain forever at this microscopic scale. Instead, cooperatives of these entities emerged to form larger-scale entities. Then, cooperatives of these emerged to form living entities of an even larger scale. This process repeated itself many times in a step-wise fashion. Now we have cooperative organisations on the scale of continents, e.g., nation-states.

However, at every step in this long climb towards ever-increasing integration, evolution had to find a way to overcome the competitive dynamics that undermined the emergence of larger-scale cooperation. Evolution needed to overcome the barriers to the emergence of cooperation which Richard Dawkins and others have identified so clearly. It needed to find a way to combine the entities at each level into the cooperative organisations that became the larger-scale entities at the next level.

However, unlike most fisheries managers, evolution has found ways to overcome these kinds of challenges. It discovered ways to solve the universal problem that otherwise impedes the evolution of cooperative organisation, no matter how beneficial the cooperation.

For example, evolution has produced our bodies, which are each cooperative organisations of trillions of cells. Every one of our cells spends its existence pursuing its own immediate cellular interests. Yet somehow, these self-centered actions combine to produce our thinking, our speech, and our digestion.

I became a fisheries manager because of my obsession with fishing. But doing so drove me to gain a far greater understanding of why the trajectory of the evolution of life on Earth takes the shape that it does. Gradually, I began to see why the forms of organisation at all levels of living processes have common dynamical architectures.

I realized that they took this form because it solved the fundamental challenge of how to organise self-interested entities into a cooperative. Consequently, this architecture manifests in all living organisations, from cells to multicellular organisms to human corporations and nation-states, and so on. I began to see how these forms of organisation operated to ensure that the pursuit by individuals of their own immediate interests caused them to act in the interests of the organisation as a whole.

Once I saw this, it became obvious why natural section drove evolution towards the step-wise emergence of cooperative organisation of ever-increasing scale: once evolution overcomes the barrier to cooperation at a particular level, larger-scale cooperatives can emerge successfully at that level. These cooperatives have the potential to outcompete individual entities that continue to live a solitary existence and that do not join cooperatives. As we see continually at the human level, cooperative teams have the potential to outcompete isolated individuals acting alone.

For example, cooperatives can take advantage of the adaptive benefits that can flow from divisions of labour, synergies, specialisation, and other forms of differentiation. Furthermore, the larger the scale of a cooperative, the greater its command over resources and the greater its power in interactions with others and with its environment.

Provided evolution finds a way to overcome cooperation barriers, cooperators will inherit the Earth.

I was beginning to see the answers to one of the key challenges I had set myself in my evolutionary thinking: to identify the causal mechanisms that drove the trajectory of evolution towards the increasing integration of living processes, as identified by Teilhard de Chardin. The answers that I stumbled upon also provided an understanding of why living processes are organised the way they are. These realizations eventually produced a general theory of living organisation. This general theory could explain the dynamical architecture that characterises living organisation, and why it emerged in the form it did.

However, my time as a fisheries manager also helped me to achieve some goals that I had formed much earlier in my life. As I indicated earlier, when I finished year 12 as a 16-year-old and thought about what course I would do at university, I decided on Marine Biology and Zoology. In my simple-minded naivety, I thought that if I had the misfortune to have to work for a living, those career choices would give me as much time as possible to go fishing. When I became a fisheries manager whose responsibilities extended to the tropics of far northern Australia, this early dream seemed well on the way to being fulfilled. Any time I wanted to escape the inland cold of a Canberra winter, I could organise a meeting that would take me to the tropics of Cairns, the Torres Strait, or PNG.

The high point of living this dream came when our PNG fisheries counterparts organised a meeting on Loloata, a tiny coral island that lies off the PNG coast near Port Moresby. We convened our meeting in an open-sided building with a thatched roof that was next to the island's jetty. Before the meeting, I caught a 12-inch pike that I cast out as a live bait under a float. We began the meeting on the basis that when the sound of my reel's ratchet indicated that the bait had been taken, the meeting would be adjourned while I played whatever fish had taken the bait.

I could not imagine it getting any better than that.

However, given my commitment to evolutionary theorizing, the price of continuing as a fisheries manager in the long term was too high. To do the job properly, I had to work long hours. In order to ensure that appropriate management was put in place for each of my fisheries, I had to spend a lot of time manipulating the disparate interests. I was not prepared to settle for anything less than doing the best that I could, irrespective of the demands it placed on my time. Eventually, I came to the conclusion that, like teaching and union work, staying in the job long term would not give me sufficient time to do my evolutionary work.

As I began to realize that continuing to work as a fisheries manager would not give me the spare time I needed, I looked for alternatives that would. Initially, this sent me down a path that seemed to bear little relationship to my evolutionary work. I set out to invent and patent a new kind of fishing lure. However, what initially looked like a rather desperate detour, produced an important development in my cognitive capacity. Again, my development was propelled by pressing necessity.

Over the years, I had noticed a fundamental limitation in the types of fishing lures that were available. What Americans call crankbaits are arguably the most effective fishing lures yet developed. These are hard-bodied, baitfish-imitating lures that wobble strongly when retrieved through the water. However, they were effective only if they could be put in front of fish. But this is where they often fell short in practice. The problem was that if they were combined with sufficient weight to enable them to be cast long distances or to be fished in deep water, it would kill their action. The extra weight would seriously dampen their capacity to wobble.

So, when I was looking for a way to earn enough money to fund my theorizing and writing, I thought that I should invent a new class of fishing lures that overcame these limitations. Eventually, I was successful. I found a way to integrate weight with a wobbling crankbait in such a fashion that the wobbling was preserved and, in some cases, actually enhanced.

I made some prototypes, and they worked well. Then, to ensure that I could make money from the invention, I had to patent it. Otherwise, it could be copied freely. I applied for international patents and proceeded first with getting a US patent, given that the US is the largest market for fishing lures.

The trick in drafting a patent application is defining the invention in such a comprehensive fashion that it is impossible for manufacturers to get around the patent. The goal is to ensure that all possible worthwhile instantiations of the invention are covered by the patent.

Working out how to do this required me to envisage all possible ways to construct my invention. I needed to form a mental model in my head, and then vary the model by changing key components of the lure in order to identify all the ways in which it could be realized effectively.

At first, I was not very good at this. I had many years of experience building and playing with abstract mental models in my head. Abstract thinking was my bread and butter. But here, I needed to build concrete mental models. I had very little practice in doing this. As a young boy, I had next to no interest in playing with mechanical objects or making anything concrete. This was where my brother excelled. As my parents put it, he was 'good with his hands', while I was 'good with my head'.

I found that my attempts to build concrete mental models and be aware of my mental processes when doing so, were similar to when I first began to try to become aware of my abstract, problem-solving techniques in year 11 at school. I looked inside to see my mental processes, but initially I could not see much at all.

I remember when the breakthrough in my concrete thinking occurred. I was on a plane flying from Sydney to Norfolk Island to consult with the Islanders about establishing the first offshore fisheries management arrangements for the Island. Suddenly, I could see complete instantiations of my fishing lure in my mind. I could change particular components in the design and morph them into different forms. I was able to observe the effects of these changes on the functioning of the lure.

I completed the patent applications. The patent was granted after I negotiated a few amendments with the US Patent Office.

The good news ended there. I made several prototypes and sent them to major fishing lure manufacturers. When the Internet became available, I created a web page to promote the lure. I sent a prototype to Vic McCristal, Australia's foremost fishing writer. He got its significance. Vic described the lure as a complete breakthrough in fishing lure design and wrote about it in one of his articles.

But Vic was the only one. I reached the point where the time I spent searching for a manufacturer seriously restricted the time I could devote to my evolutionary work. I decided not to proceed further.

However, as now seems common when I look back at my life, the challenges that I encountered with the fishing lure had important unforeseen consequences for my development. I now had the ability to build, test and recursively improve concrete constructions in my head. Up to that point, I had never been a handyman. But now, I was able to save a lot of money by designing and building fences, gates, and other structures for our new home in Canberra. But more importantly for me, it broadened my ability to build and to recursively improve mental models for my evolutionary thinking.

11.

Pretending to Work for the Government

After three years as a fisheries manager, I applied for and got a policy development job back in the Department of Industrial Relations. As a class 11 Section Head, I had my own office and four or five staff. The work was not very stressful or demanding. Most of the time, I found that I could easily complete the relevant tasks well within the minimum number of hours required for attendance each week.

Nevertheless, as a perfectionist, I still tried to ensure that the issues I worked on were handled fairly and effectively. However, as a bureaucrat behind a desk in Canberra working on high-level industrial relations policy issues, I was largely insulated from the impacts of the policies I helped develop.

This was quite different from my previous work in fisheries management, teaching, and unions. If, due to the incompetence of the system or of individuals, I could not ensure that appropriate industrial relations policies were adopted, I did not have to face those who suffered the consequences. Bad industrial relations policies could do far more harm than poor fisheries management or teaching. But I used to tell people that high-level policy development was much less stressful than my previous work. It was more like dropping atomic bombs on Hiroshima. You did not have to experience first-hand the harm it caused.

The pressure I put on myself to make tangible progress with my evolutionary thinking intensified as the years passed. It was 1990, and I was fast approaching 40 when I recommenced working in the Department of Industrial Relations. My reading and thinking were progressing significantly, but I had not yet developed anything that could be published in a peer-reviewed scientific journal.

To further enhance my capacity for 'big picture' thinking, I decided to begin practicing a form of meditation. My reading had suggested that meditation could be particularly useful for promoting creative thinking. I was also attracted to the possibility that meditation could help to further disrupt my analytical/rational cognition and, at the same time, give me greater access to intuitive forms of knowing. Both are important for enabling 'big picture' thinking.

I also thought that meditation might be able to enhance further my ability to see my own thinking as object. The capacity to examine my own thinking was central to my work on recursively improving my cognitive abilities. It enabled me to see where my current thinking was inadequate and to improve it intentionally. The efforts that I had made since I was in school to recursively improve my thinking had helped me to become aware of the thinking strategies that I was using and to see how they might be enhanced. However, I hoped that meditation could further develop this capacity, particularly in relation to thinking about complex phenomena.

Finally, as I mentioned earlier, I had used the Gurdjieffian technique of self-observation to some extent to free myself from cravings for cigarettes and overeating. As I read more widely, I learned that self-observation was very similar to the meditative practice of mindfulness. This was useful because practices for achieving mindfulness were far more widely known and accessible than Gurdjieff's techniques.

I should emphasise that I did not take up meditation for the purpose of 'spiritual development'. Instead, I was solely interested in exploring its potential to enhance my capacity to be effective in this world, in the midst of ordinary life. I did not have a spiritual bone in my body.

I was attracted to Gurdjieff's philosophy because it seemed to advocate self-mastery, and talked about some tools for achieving this. It was focused on developing capacities that would enable me to be more effective in achieving my goals. In particular, a central aim of his practices was to 'wake up' in the midst of ordinary life. Gurdjieff's approach was not interested in meditation that was confined to the meditation cushion, and that explored different states of consciousness primarily on the cushion. It certainly was not directed at withdrawal from the world. Rather, his practices aimed to achieve states of consciousness that could increase one's effectiveness and agency in life.

Sometimes the Gurdjieffian approach has been referred to as a 'left-hand path'. It contrasts with 'right-hand paths' that encompass traditional spiritual approaches. The goal of right-hand paths is 'absorption in the absolute', and their maxim is 'thy will be done'. The goal of left-hand paths is 'self-mastery', and their maxim is 'my will be done'.

I read widely in the spiritual literature, including the work of great synthesisers like Ken Wilber. But my primary goal was always to collect techniques and practices that could be adapted to achieve self-mastery and enhanced agency in this world. However, the overwhelming majority of material and practices that I found related to the right-hand path.

But again, necessity was the mother of invention. I set out to develop an understanding of the principles that underlay the practices of the great contemplative and spiritual traditions. My goal was to develop comprehensive models that explained how the practices worked. Once I had achieved this sufficiently, I could then use these models to explore how the practices could be modified and enhanced.

In particular, I intended to use the models to search for modifications that would enhance their ability to produce self-mastery rather than only transcendence. Of course, this task of building meta-models of meditation, mindfulness, self-observation, and awakening fitted perfectly my central interest in recursive self-improvement and its role in the future evolution of humanity.

I was used to building mental models of my own mental processes. As I have described, I did so to identify modifications to my thought processes that would enhance my thinking and problem-solving. This general approach is broadly analogous to how science has driven technological development.

First, science develops a model of particular natural processes. These models can then be used to predict what happens when components in the model are changed. This enables models to identify changes and interventions in the real world that will produce outcomes that better serve particular human needs. These changes can include the making of tools and, eventually, machines. They provide the foundations for technological advancement.

In effect, I was setting out to technologize the practices and methods of spiritual development. The goal was to develop psycho-technologies that could produce self-mastery, including by enhancing human cognition.

The capacity for an individual to be 'non-attached' to thoughts provides a concrete example of this technologizing process. Several of the spiritual traditions teach meditation-like practices that are claimed to assist an individual to become 'non-attached' or 'dis-identified' from thoughts that arise in their minds. These practices include mindfulness and also methods that involve the individual surrendering to whatever arises in the moment (surrender may be facilitated by, for example, a belief that if one hands over control of one's life to Jesus, one will be happier and more content here as well as in an after-life).

In general, these practices seem to work successfully. Significantly, however, they work irrespective of the supernatural or mystical explanations given by any particular tradition to explain why their practices are effective.

The strategy I am outlining here is to develop a scientific understanding of why the practices developed by the traditions can be effective. In the specific case I am exploring, the goal is to develop scientific models that explain how particular practices enable individuals to dis-identify from their thinking. Once these specific explanatory models have been developed, they can then be used to develop various forms of self-mastery.

For example, they can be used to enable an individual to detach from debilitating worrying that could otherwise undermine their effectiveness in work and other activities. Once they can achieve non-attachment to their thinking at will, they can then passively watch any worrying thoughts as they arise and then dissipate, without clinging to or worrying about them. With this capacity, a person can be as happy and un-troubled as a born-again Christian, but without having to surrender their life to Jesus.

Practices that hitherto were explained by spiritual traditions as being the product of supernatural causes can then be shorn of these explanations. They can be replaced with scientific models that can be used to develop practices that are more effective.

I began my meditation practice with a standard mindfulness approach. I would attempt to rest my attention continually on the sensations of my breathing. When I found my attention had moved back to thinking, I would gently return attention to my breath, and so on, repeatedly for 20 minutes. This practice would still my mind to some extent. However, I found that giving attention to the sensations of my breathing tended to disturb my mind, apparently due to the dynamic nature of breathing.

So, I changed to a form of concentration meditation. I would rest my attention on the head of a small ornamental doll that was on a sideboard near where I sat. At times, I would do the same with a dot on the wall. Using a similar approach to the mindfulness practice, when I found that my attention had wandered, I would return it to the object of concentration, and so on, repeatedly. This approach seemed to strengthen my ability to notice more quickly whenever my attention had become re-embedded in thought, to disengage from this embeddedness, and to remain dis-embedded from thought for longer periods.

Within about a week of starting concentration meditation, I found that after 15 or so minutes of the practice, my mind would cease to be disturbed, and I would enter a state in which I could rest attention more or less continually on the object of attention. It is difficult to convey the nature of this state in words, but it was as if the mind became solid and undisturbed. A friend describes it as 'the lake freezing over'. Another description that captures a central aspect of how I experience the state is a phrase I often heard when my parents took me to church when I was young: "The peace that passes all understanding." I have also heard it said aptly that entering this state is always surprising—it is so different from ordinary experience.

Of course, there is no sequence of words that I can write down here that will produce this state in you and enable you to experience it. However, if you do the necessary work on yourself to enter this state, you will know precisely what these various descriptors are attempting to point to.

Initially, ensuring that I did the 20 minutes of practice each and every day was a challenge. Frequently, other activities intervened. The strategy that changed this was to get out of bed 20 minutes earlier than I did previously. This became a habit that always gave me the extra minutes every morning that I could use to meditate. I also found that it was easy and productive to perform additional meditation when I was on a plane flight, waiting for a doctor's appointment, eating at a food court, on my lunch break, waiting for public transport, and so on.

Meditation in public places also seemed to be consistent with my goal of awakening in the midst of ordinary life, rather than just on the meditation cushion. One of the key reasons for meditating alone in a quiet, darkened room is to avoid distractions that might otherwise capture your attention and disrupt the concentration that leads to the stilling of the mind. In the midst of ordinary life, potential distractions are legion. To awaken in ordinary life, it therefore seemed necessary to progressively strengthen the capacity to stay in a meditative state in the face of distractions.

Furthermore, it seemed to me that I should also progressively develop the ability to meditate with my eyes open. Again, I hoped that this would strengthen my ability to achieve meditative-like states when I was out and about in the world, despite all the distractions.

However, despite extensive searching, I could not find specific practices designed to facilitate the transition from the meditation cushion to awakening in the midst of ordinary life. Gurdjieff had specifically advocated this, but as far as I could determine, his writings do not set out in detail particular practices that would scaffold this transition.

In general, spiritual traditions have not even established this transition as a major goal, let alone developed practices to facilitate it. They seem to have always been strongly attached to the right-hand path.

Nevertheless, for many years I persevered with the form of meditation practice that I have outlined. I cannot be certain that it actually enhanced my evolutionary thinking. At the time, there were too many variables to assess what changes caused particular outcomes.

However, at the beginning of the 1990s, while working in the Department of Industrial Relations, I began to make progress in my 'big picture' evolutionary thinking. My meditation practice might have helped with this and enabled me to deal with work and family pressures with greater equanimity. But in any event, it became an enjoyable habit, and I continued.

As had become increasingly evident throughout my life, challenging work demands tended to propel my further development. In 1991, I was installed as the acting head of the Employment Conditions Policy Branch in which I worked. This position was at Senior Executive Level One in the Australian Public Service and put me in charge of four Sections, including the one I previously headed. I was appointed temporarily but indefinitely, with the expectation that I would eventually be appointed permanently, if all went well.

But for me, it did not go well. I had been making progress with my evolutionary thinking, but this stopped as I began working long hours as a Branch Head. Previously, I was responsible for the work of one section, and now there were four. Furthermore, I soon discovered that I could not rely on any of the four Section Heads to produce work at a level that I considered to be competent. As a Section Head, I found that I could easily work around the lack of competence of my staff. I could quickly rewrite the Ministerial briefs that they produced. In order to minimize the rewriting required, I could even dictate the intellectual content of briefs to them in detail so that they did not have to think deeply or originally about the relevant issues.

But as a Branch Head, I could not do this easily within normal working hours for four Sections. And none of the Section Heads were up to my standard. At first, this greatly surprised me. I began to realize that this was not because my section Heads were exceptionally bad. Rather, it was because I did not have much experience of other people's ability to do challenging and intellectually demanding work. In general, I just assumed that most people could do what I did.

For most of my working life, I had not been in charge of high-level staff. In my work for unions, I had generally got myself into a position where I was essentially a one-man band. I tended to lead others, whether they were working for my union or other unions. I would develop strategies, and they were happy to follow my plans.

Even at school, although I did much better at exams than nearly everyone else, I did not think much about why this was so. When I did, I thought that it was not due to any fundamental difference in the level of ability between myself and others. I tended just to assume that others could do what I did if they put their mind to it. I did not believe that I was exceptional in any sense. When others in school, university, or work treated me as if I were, I would quickly brush it off as them being nice. It never sunk in. If anything, I felt embarrassed being singled out. Being smart was not something I cared about or valued.

In the few years I spent as a Section Head in fisheries management and policy development, I thought it was just bad luck that the staff I inherited were often not very competent. It was not until I became a Branch Head that I realized that very few people had the intellectual capacity to do high-quality policy development work or to strategize effectively in complex circumstances.

I realized quickly that if I were to do the Branch Head job to anywhere near my standards, I would have to work long hours on intellectually demanding tasks. This would seriously impede my ability to spend 'quality time' on my evolutionary theorizing. And I did not need the extra money. I had always made sure that I could afford my housing mortgage repayments even if my income dropped significantly.

After four or so months in the position, I told my superiors that the Branch Head level was beyond my capabilities and did not suit my personality. I asked to be returned to the Section Head level.

This greatly surprised my superiors. In their view, I was doing the job extremely well. And it was very rare that a person rejected promotion to the Branch Head level and the substantial increase in salary and provision of a car that went with it. For many public servants, getting there was their lifetime ambition.

But they had no choice but to return me to the Section Head level and bring in someone from elsewhere to fill the Branch Head position.

In the short term, the person they appointed was an unfortunate choice for me. He was a leading example of the kind of person who was increasingly getting Senior Executive Service jobs. My new boss was a compulsive 'can do' man. If the Minister or his other superiors gave any hint that they might want something, he would promise to deliver it urgently. This was despite the fact that often his superiors had not thought through the issue, nor undertaken any kind of cost/benefit analysis of whether it was worth the resources, nor made any assessment of how long it would take, and so on.

They would say "jump", and he would always say "how high?" He would return to his Sections, tell them what had to be done, and set a tight deadline. It was not negotiable or subject to reality-testing. No discussion was permitted. Worse still, sometimes projects that he had instructed us to treat as high priority would be dropped by his superiors after a few days, and he would forget to tell us as we continued to slave away.

So I had jumped from the frying pan into the fire. My new Branch Head used his staff like a World War 1 General. He would order his troops out of the trenches to charge the enemy lines, unprotected in the face of withering fire. And if his staff and his strategy failed, he would blame it on his staff.

By this time, I had begun to make real progress in my evolutionary theorizing. Having my Branch Head continually get in the way of further progress was a major source of frustration. But in the longer term, this proved to be the best thing that could have happened to me at the time.

Up until these events, I took responsibility for whatever my job required. I did what I could to fulfil the requirements, without question. This did not mean that I did what I was told by my superiors. When they gave me instructions that conflicted with the wider responsibilities of my work, I tried to convince them that alternative action should be taken. I would manipulate my superiors and those I interacted with to ensure that the wider responsibilities of my job were achieved to the maximum. My overriding goal in my work was to achieve the objectives of my job as I understood them, not as my superiors understood them, particularly if my superiors were incompetent. First and foremost, I was loyal to the responsibilities of whatever job I was being paid to do.

I consciously and intentionally dropped this loyalty in 1992. I never picked it up again.

This was due to the development of my evolutionary thinking as well as to the conflicts between it and work demands. As my 'big picture' thinking had progressed, I saw that there was no moral force behind our society's requirements in relation to paid work. The basic requirement was that in order to get sufficient money to sustain yourself and family, you had to sell control over your consciousness and your being for at least 40 hours a week.

Nothing in our current legal and economic systems guarantees that you will be able to spend these 40 hours using your talents in whatever way benefits society the most. It was clear to me that by far the best way I could contribute to society was by developing new theories of evolution. In the coming pages, I will substantiate this claim in detail. However, our current social and economic systems did not give me the opportunity to earn a living doing this. I could not find any possibilities in academia or elsewhere that would enable me to do so.

Our economic system currently offers enormous rewards for anyone who, for example, develops a better-tasting ice cream, a computer game that totally absorbs the attention of players for hours on end, or a more effective weapon for killing people. But it offers no immediate rewards for someone who spends years developing a new 'big picture' theory of evolution that can help guide humanity to reorganise our societies and our psychological processes in ways that benefit us all.

Increasingly, I realized that I had no moral obligation to spend my life performing work that wastes my talents and fails to make the best use of my abilities. For someone like me, work steals my consciousness and fritters it away on trivialities. This conclusion was valid whether my superiors were competent or not. But I was propelled to it much more quickly by incompetent superiors who wasted my potential to perform useful work even in industrial relations policy development.

In 1992, I decided that henceforth, my overriding goal in my paid work was no longer to fulfill the responsibilities of the job. Instead, it was the same as my central objective outside work: to develop a new theory of evolution and to identify its implications for humanity, both individually and collectively. This led me to the decision to manipulate my superiors, those around me, and the system in which I was embedded to ensure that I would have enough time to devote to my evolutionary theorizing.

I no longer just accepted that my superiors could require me to do whatever work they decided. Now, I used my strategizing abilities and persuasive capacities to control the kind and amount of work I was given.

This did not mean that I was always avoiding work. For example, I would willingly take on projects where I could control what was done. I would happily volunteer for major projects where I would be the only person who knew fully what was being done, what meetings with employer and union representatives should be arranged, how long meetings would take, when interstate travel was needed, and so on.

In particular, I manipulated circumstances so that I became responsible for a number of major test cases that were conducted by State and Commonwealth arbitration tribunals. In these cases, I took responsibility for drafting the Government's submissions and for presenting them to the tribunal hearings as the Government's advocate. These cases could take more than 6 months of intensive work, and as the only person in the Department who knew in detail what was required, I was well-positioned to control what my superiors expected me to do.

For well over a decade, I travelled around Australia representing the Government in negotiations with unions and employer groups and as the Government's advocate in significant cases before industrial tribunals. I would confidently and articulately put forward the Government's position on the relevant issues.

Generally, my Section and I wrote the submissions that I put to the industrial tribunals. But often, I did not get the details approved by my Departmental superiors or by the Minister's Office, which had ultimate responsibility (and power). To do so would have multiplied considerably the amount of work that had to be done. The chain of command above me would have been unable to resist the temptation to require significant rewrites, commission additional research, include further arguments, and so on. Generally, all this additional work would have detracted from the quality and effectiveness of the submissions. None of my superiors or the Minister's office had the experience or intellectual capacity to improve my submissions.

Often, I was able to avoid getting specific approval for the details by giving my Departmental superiors the impression that the details had already been approved by the Minister's Office or by other Departments that were involved. I would often give the impression to the Minister's Office that the detail had already been approved by the Departmental hierarchy or by the staff of other Ministers who were involved.

I knew that I was taking a risk. But I judged that I could get away easily with being caught out at least once. If this happened, I planned to say that I thought that I was operating within the bounds of existing policy positions that were already well-established and that reflected the Government's position in detail. I planned to then apologise profusely and abjectly. I knew that I could get away with this the first time without disciplinary action. From then on, I would have to get all the appropriate clearances. However, it never happened a first time.

But this did not mean at all that I would refuse to do work that I was given. Or that I would do poorly whatever work I could not avoid. I continued to perform at a very high level any work that I could not avoid.

In fact, given my extensive experience in all aspects of industrial relations and the years I had spent recursively improving my cognitive capacities, I did not do the work just a bit better than anyone else in the Department. Rather, I could achieve outcomes that were a quantum level above what anyone else could have. I could see things they could not. In the part-time hours I devoted to my paid work, I could do what others could never do, no matter how much time they spent working.

This is reflected in the fact that in my annual performance reviews, I was generally rated at the highest possible level (outstanding). Furthermore, for my efforts in 1999 in persuading a Full Bench of the Australian Conciliation and Arbitration Commission to support the retention of Australia's system of junior rates of pay, the Minister Peter Reith thanked me in a speech to the Parliament for my efforts. Later, he rewarded me with a three-week trip to Geneva as one of the two Australian Government delegates to the International Labour Organisation Annual Conference in 2000.

Shortly before I retired from paid work in 2007, the Head of the Department gave me his Outstanding Achievement Award in recognition of my work on the Redundancy Test Case that was conducted by a Full Bench of the Australian Commission.

Nevertheless, after this fundamental transition in my approach to paid work in 1992, I used to say that although I was employed by the Government, I did not work for it. Instead, I worked for myself and my evolutionary goals. After 1992, I only worked for the Government part-time, although they continued to pay me full-time.

Ultimately, however, I knew that my evolutionary work would make a much greater contribution to humanity (including to the Australian taxpayer), than anything I could ever do in my paid employment.

HUMAN SUPERINTELLIGENCE

12.

Evolutionary Progress

From 1992 onwards, my evolutionary work progressed rapidly.

I still remember a breakthrough that occurred with a rush of vivid images and insights. I saw in an instant how key issues that I had wrestled with for many years could be resolved. At the time, I was standing in the family room of our home in Canberra. It was a very cold morning in mid-winter, but the sky was cloudless blue, and the sun was streaming in through the north-facing windows. The floor-to-ceiling windows had been designed to capture the light and warmth of the sun as it moved northward and lower with the onset of winter.

Of course, this rush of insights did not come out of nowhere. For years, I had put a lot of hard intellectual work into developing mental models of various 'big picture' evolutionary processes. I had many specific insights into the nature and operation of these processes. But what my epiphanous experience provided was the sudden ability to see commonalities in the deeper, dynamical structures that underlay these phenomena. In an instant, I realized that processes and systems that I had previously seen as separate were united at a higher level of abstraction and shared common dynamical architectures. In a rush, I suddenly became aware of connections and unifying insights that I had not seen before.

In particular, the flash of insights produced the realization that two important forms of organisation were identical, even though they differed enormously in scale. From a very abstract perspective, I saw that a planetary society managed by a global system of governance had the exact same organisational architecture as a single cell managed by the cell's genetic apparatus. As above, so below.

This central realization was accompanied by several insights. Foremost amongst these was that both global governance and the genetic apparatus performed the same functions within their respective organisations: they both used their power to support cooperation and to deter free-riding that would otherwise undermine cooperation. Both managed/governed societies: the global system was a self-producing society of humans, technology, and other living processes; the cell was a complex society of self-producing molecular processes and technologies.

I saw immediately that these common dynamical architectures arose in both cases to solve exactly the same kind of evolutionary challenge. Furthermore, I realized that this understanding enabled me to identify the key mechanisms missing from Teilhard de Chardin's hypothesis that evolution embodied a trajectory toward increasing integration. As I have mentioned, Teilhard had failed to demonstrate how complex cooperation evolved.

Up until this time, evolutionary science had done a very thorough job of demonstrating how difficult it was for cooperation to emerge, no matter how beneficial the cooperation might be. It was not that there was any doubt that if cooperation could emerge and persist, it could be favoured by natural selection.

It is obvious at the human level that organised cooperative groups generally have the potential to outcompete isolated individuals who act alone. It is equally obvious that once the evolutionary process discovers a way to reproduce cooperative organisation, it can flourish and prevail over non-cooperators. The emergence of multicellular organisms from single cells, and their subsequent differentiation into many forms that spread across the planet, is a powerful example of the potential adaptive benefits of cooperation.

As I have mentioned previously, the fundamental reason why the evolution of cooperation tends to be impeded is easy to see: a cooperative organism invests some of its resources in cooperating with others, rather than using those resources for its own needs (for example, an organism might share food with others when it has some, but the others have little. This kind of behaviour might be very valuable to those who get the shared food, who might otherwise perish).

However, if cooperators invest some of their resources in helping others, they will end up worse off than non-cooperators. In contrast, organisms that do not cooperate will refrain from investing any of their resources in cooperating with others, but will accept resources from cooperators (in this example, an organism that never shares its food cooperatively with others will do better than those that do, because it will retain its food and also get the benefit of food shared by cooperators).

In short, cooperators will not emerge and persist unless they capture sufficient of the benefits of the cooperation that they produce.

These difficulties were a major problem for evolutionary science. First, there was a strong demand for theories that could plausibly explain the many cases of cooperative behaviour found in nature. There was an obvious need to account for the cooperation found within, for example, ant colonies, baboon troops, and human tribes. An even greater challenge was to explain the numerous cases where very complex cooperative organisations had emerged in what came to be known as major evolutionary transitions (e.g., the emergence of complex eukaryote cells from cooperative communities of simple cells; of multicellular organisms from groups of eukaryote cells; and of hierarchical human societies from tribes).

Second, there was a large market both within evolutionary science and beyond for an evolutionary narrative that emphasised the benefits of cooperation, at least to some extent. Few wanted a story that suggested that evolution always favours those who relentlessly pursue their own immediate interests, including at the expense of others.

As a result of these demands, evolution science is littered with failed attempts to develop theories that attempt to explain the diversity of cooperation found in nature.

Early on, group selection was proposed as the main contender. The basic idea was that cooperative groups could outcompete groups that did not contain cooperators. Because this model predicted that evolution could favour cooperation in particular circumstances, and because many hoped that this reflected how the world was organised, group selection theory was not subject to strong criticism for some time.

But in the latter half of the 1960s, group selection was rejected emphatically for a reason that looks obvious in retrospect. Within a cooperative group, any non-cooperators that received the benefits of cooperation but did not contribute to the group cooperatively, would be favoured by natural selection within the group. These free-riders would outcompete cooperators and undermine cooperation within the group. Within a group, cooperators would tend to be selected out of existence.

This criticism of group selection was considered a knock-out punch for some while. Other explanations for the emergence of cooperation were developed, but they had limited applicability. They failed to account for the complex cooperative organisation that arises in the major cooperative evolutionary transitions.

These limited explanations include kin selection theory. It argued that the genes that predisposed an organism to cooperate could prevail by preferentially directing the cooperation towards relatives. This was because relatives were more likely to also carry the cooperator gene, and less likely to carry non-cooperator genes. As a result, where kin selection operated effectively, cooperator genes were more likely to capture the effects of beneficial cooperation, and non-cooperator genes were less likely to be the beneficiaries of cooperation.

A further example of a theory of the evolution of cooperation that seemed plausible but only in limited circumstances was reciprocal altruism. This mechanism was based on the notion that cooperators who cooperated only with other individuals who also cooperated (e.g., reciprocators), could capture sufficient of the benefits of cooperation, and exclude non-cooperators from the benefits.

A large academic industry grew up around the development of mathematical models of these and other mechanisms. Game-theoretic models were particularly popular. Modellers searched far and wide for particular circumstances in which cooperators and pro-social behaviour would flourish, at least in their models. This work tended to produce rigorous mathematical analyses of circumstances that bore little relationship to complex reality. Journals that published papers on evolutionary theory were filled with these kinds of junk papers that were soon forgotten.

In general, the only valuable outcome achieved by this considerable investment in research was to demonstrate how difficult it is for complex cooperation to be favoured by evolution.

However, it was not long before group selection made a comeback. Mathematical models suggested that in specific, limited circumstances, selection that operated at the level of competition between groups could prevail over selection that operated on individuals within groups. These models were based on the earlier idea that a group of cooperators could outcompete groups with no cooperators due to the adaptive benefits of cooperation.

However, the new models went on to suggest that this advantage at the group level could be sufficiently strong to overcome the disadvantage suffered by cooperators within groups (as we have discussed, cooperators tend to be outcompeted by free-riding non-cooperators within groups). The boost that cooperators would get when they were in a group of other cooperators could be strong enough to outweigh the disadvantage they suffered within groups. I will refer to this as Wilsonian group selection for reasons that will soon become clear.

Since the rebirth of group selection, it has gained acceptance among many evolutionary scientists. It is seen as capable of explaining the examples of cooperation observed in nature, including the complex cooperation that manifests in major evolutionary transitions. The leading proponent of this rehabilitation and subsequent ascendancy of group selection is America's foremost evolutionary scientist, David Sloan Wilson.

However, it seems to have been largely forgotten that current group selection models have only explained cooperation in very limited circumstances. In part, this collective amnesia seems to be due to the demand for positive evolutionary stories that support pro-social values. Certainly, David Sloan Wilson's recent books seem to cater to this market intentionally.

However, perhaps a better explanation is that no plausible alternative has yet attracted widespread support. No other theory has yet been accepted as explaining the extensive existence of cooperation in nature, particularly in complex forms such as in the major transitions. This is despite widespread acceptance that cooperation has the potential to be selectively advantageous once it can arise in a stable form. The frequent existence of complex cooperation in animals and humans suggests that an evolutionary mechanism that produces cooperation must exist. In these circumstances, if there is only one theory that appears to be even somewhat credible, it is highly likely to be taken to be a valid explanation. Hand-waving and wishful thinking can go a long way when circumstances demand an explanation and no viable alternative exists. The only theory that is at least partly plausible tends to be taken as the explanation.

The limitations of the group selection mechanism are most obvious at the level of human societies and human organisations. Examples at the human level are particularly useful because we have a lot of direct experience as participants in these kinds of cooperative groups. This provides us with an intuitive understanding of their structure and functioning.

For this reason, I will sketch some examples at the human level that illustrate the limitations of group selection and that also identify the form of organisation that enables the emergence of complex cooperation. I will go on to show how this form of organisation not only provides a solution to the cooperation problem at the human level, but also provides the mechanism that overcame the cooperation problem at the molecular level, enabling the emergence of life itself.

This mechanism subsequently enabled all further major cooperative evolutionary transitions. As such, it is the mechanism that I saw in a flash of insight in my family room in Canberra one sunny winter's morning. It is capable of organising complex cooperation within a cell and also within a global society. For reasons that will become obvious, I refer to the set of ideas and models that explain this mechanism as Evolutionary Management Theory.

For my first example, I will consider the architecture of the form of social organisation that constitutes a typical modern Western democracy. A central feature of such a society is a government that has the power to establish laws and institutions that define certain classes of behaviour as illegal. The government also has the power to establish a legal system, a police force, and related arrangements that enforce these laws.

These systems of laws typically operate to ensure that much behaviour that would otherwise undermine cooperation between citizens is punished and therefore not in the interests of citizens. These include laws that punish behaviour that would undermine a market economy. For example, the laws punish murder, theft, breaches of contracts, free-riders who do not pay applicable taxes, and so on.

The government also taxes citizens in order to fund activities that are generally claimed to benefit the society as a whole. For example, the government funds an education system, a defence force, its legislative and policing functions, and so on.

In short, modern Western governments have the power to control and manage the society from which they emerge. Ideally, they use this power to punish citizens who act against the interests of the society, and to tax citizens in order to fund behaviours that contribute to furthering the interests of society. Ideally, this use of power by governments tends to align the interests of citizens (and of organisations of citizens, such as corporations) with the interests of the society as a whole.

To the extent that the government achieves this, it will overcome the cooperation barrier. In principle, a government can overcome the barrier by rewarding those who cooperate and by punishing free-riders and cheaters who would otherwise undermine cooperation. The result is the complex cooperation that constitutes modern societies.

Of course, numerous examples exist in which the power of governance is not used primarily to organise a cooperative society. The systemic architecture that I have identified is also found in totalitarian human societies, including those ruled by dictators. It also featured in earlier human societies governed by kings, emperors, and so on.

This same systemic architecture is also typically found in complex organisations within modern societies. For example, corporations are governed by a Board and CEO. They can use the payment of salaries and other incentives to make it in the interests of employees to cooperate in the interests of the corporation as a whole. They can also use their right to fire employees to punish those who act in ways that undermine the corporation's interests.

In general, most complex human societies throughout human evolution have been organised by powerful external management/governance. Its powers to reward and punish tend to have been used to align the interests of the governed with the interests of the managers/governors. In these circumstances, when citizens pursue their own interests, they will tend to serve the interests of the managers/governors.

When the interests of management are aligned with the interests of the society, management will tend to align the interests of the managed with those of the society. In these circumstances, when citizens pursue their own interests, they will tend to serve the interests of the society as a whole.

What is common to all these societies is that the source of power and governance is external to the individuals who are managed. When we are members of such a society, we readily experience the governance as being external to us and other citizens.

However, this systemic architecture does not explain all complex, cooperative human organisation. Early human tribal societies were not hierarchical. There was no external source of power that governed the members of the tribe, and that rewarded cooperation and punished free-riders.

It is considerably more difficult to identify how it is that tribal societies are organised in such a way that cooperation between its members is enabled. As we shall see when we return to this issue below, it is also achieved by governance that acts across the tribe. But in this case the governance is internal to each member of the tribe, not external to them. There is no king or government or other powerful external ruler. Rather, the governance is internal to members of the tribe and distributed across them.

Returning to the consideration of externally-managed societies, it is necessary to explain how such societies emerged in the first place. It is easy to see that once a society is organised by appropriate external management, cooperators can easily outcompete non-cooperators within the society. As a result, complex cooperation that is supported by management can emerge and persist indefinitely. It is equally clear that managed, cooperative societies have the potential to outcompete societies that are comprised of non-cooperators. But how do management architectures emerge and evolve? And why would a manager govern a group in such a way that promotes cooperation? How could this be in its evolutionary interests?

Properly understood, the history of the transitions made by Mongols in the 12th and 13th centuries answers these questions. The Mongols began as a collection of tribes that often fought each other. From this, they eventually became the governors of China and other complex societies.

Genghis Khan led Mongol tribes to rape and pillage other tribes and societies. Typically, his warriors would ruthlessly plunder food and other resources from the societies they defeated, often destroying them. Then, they would move on to repeat the process with other societies. And so on.

However, the limitations of such a predatory strategy are that a complex society can only be raped and pillaged once in an extended period. Once its culture, governance, and infrastructure had been destroyed, it would be many years, if ever, before it could re-grow to have the complexity and wealth that it did before.

Eventually, some Mongols found a viable alternative that had the potential to overcome these limitations: instead of raping and pillaging a society only once, they could govern the society indefinitely, allowing them to continually harvest a stream of resources from the society for their own use. This is similar to the trajectory followed by humans who moved from predating wild animals to domesticating them.

It was a short step from this to governing a society in such a way that increased substantially the stream of resources it could produce.

As we have seen, this could be achieved by governing the society in a way that promotes cooperation and punishes free-riding, thieving, and cheating. For example, the Mongol governors could establish laws and enforcement systems that enabled members of the society to engage securely in economic exchanges. This, in turn, facilitated the emergence of a complex and productive economy of cooperative exchange relations. 'Good' governance had the potential to boost substantially the amount of wealth and other resources that the Mongol governors could continually extract from the society.

Such a transition established a certain coincidence of interests between the Mongol governors and the society as a whole. Just as farmers have an interest in the welfare and productivity of the animals they manage, the Mongols had an interest in the productivity and welfare of the societies they managed. Driven only by their own immediate interests, the governors were incentivized to govern the society in ways that promoted effective cooperation within it.

However, there were limits to this coincidence of interests: if the society had no competitors, the overlap of interests was not complete. The governors would be able to get away with over-exploiting the society. But to the extent that the society was in direct and on-going competition with other societies, this possibility was reduced. In a heavily competitive environment, the interests of the governors would tend to coincide more or less completely with those of the society they governed. Governors that over-exploited their society would weaken its ability to compete effectively with other societies, including in war. In these circumstances, the only way in which governors could protect and advance their own immediate interests was by advancing the interests of the society as a whole.

Of course, human history is littered with examples in which centralized power has been abused. In general, this is often because the governance of human societies is not disciplined effectively by competition between societies. Competition is often insufficient to align the interests of management with the interests of the society as a whole.

In the last few thousand years, this reduction in competitive pressures has resulted, to a significant extent, from the decline in the number of human societies. This has been driven by the increase in scale of successful societies in a finite environment. There are now fewer than 200 Nations on Earth today. This trend will be complete if a global society emerges. It will not have any competitors on this planet.

Because external, competition-induced constraints on the governors of modern societies have weakened, the development of internal mechanisms that constrain governors/management has become a major preoccupation within human societies. Democracy is a prime example of attempts to align the interests of governors/management with the interests of the society as a whole.

For similar reasons, a major focus of my subsequent work on the future evolution of human societies has been about how governance/management could be constrained appropriately. This issue cannot be avoided. The institution of effective global governance is the only feasible way in which a unified, cooperative, and sustainable planetary society can be organised. Furthermore, such a society is essential if the existential threats that face humanity are to be overcome. Fortunately, an adequate understanding of the evolutionary functions of governance/management helps to identify how it can be constrained appropriately, and how the problems and dangers that have been associated in the past with centralized governance can be overcome.[6]

It is also worth emphasising here that this account of the evolutionary emergence of the architecture of human societies relies on individuals acting only in their own immediate evolutionary interests. It does not require individuals to act altruistically. The predictions of Management Theory hold true even if managers/governors always act in their own evolutionary interests during all stages of the emergence of managed societies. The same applies to the members of a society: they are only expected to act cooperatively when the actions of a manager make it in their evolutionary interests to do so. In general, mainstream evolutionary science accepts that actions that are consistent with self-

---

[6] Stewart (2018) – see References for full details

137

interest in this way are favoured by evolutionary processes. Selection that operates at the level of the individual does not require any special, additional explanation.

In contrast to Evolutionary Management Theory, Wilsonian group selectionism has not been able to explain the emergence of complex cooperative organisation by selection processes that act at the level of individuals.

The core of Wilsonian group selection is that the disadvantage suffered by cooperators within a society can be overcome by the competitive advantage experienced by a society of cooperators, provided this advantage is sufficiently strong. However, it is obvious that complex, hierarchical human societies do not exhibit these characteristics of the dynamics of Wilsonian group selection. Within complex human societies, and within complex organisations within such societies, cooperators are manifestly not being continually outcompeted by free-riders or other non-cooperators.

We all know this from our own direct experience. We know that within complex societies, cooperative behaviour that is in the interests of the society often tends to also be in the interests of cooperators. Rather than being out-competed by free-riders, cooperation tends to be supported by the governance of the society. And free-riding, theft, cheating, and other behaviour that undermines this cooperation tend to be punished. We can all point to some exceptions, but it is certainly not the case that, as proposed by Wilsonian group selection, cooperation is generally out-competed within societies except in limited cases (e.g., where kin selection or reciprocal altruism applies).

Furthermore, Evolutionary Management Theory does not suffer from the deficiencies exhibited by attempts to extend Wilsonian group selection to explain the emergence of complex, cooperative organisation in major evolutionary transitions. These extensions to Wilsonian group selection were made because it was becoming increasingly accepted that the standard Wilsonian model could not account for these cooperative transitions. In most situations, group selection is not strong enough to outweigh the disadvantage suffered by complex cooperation within groups.

Typically, these ad hoc extensions rely on the emergence of special mechanisms that reduce competition within groups. Wilson envisages that these mechanisms would dampen the disadvantages suffered by cooperators within groups. Consequently, where these mechanisms operate, cooperators can survive and thrive within groups, and selection between groups will favour groups with the most effective cooperation. Examples of these mechanisms at the human level that are given by Wilson and others include norms and institutions.

However, extended Wilsonian group selection fails to explain convincingly how these competition-suppressing mechanisms would emerge in the first place. In particular, proponents of group selection have been unable to demonstrate that selection operating at the level of individuals can explain the emergence of these complex mechanisms.

Given this failure, Wilsonian group selectionists are left with a major challenge—they have to demonstrate that the suppression mechanisms can be established by group selection. But this creates a serious problem for their extended theory: they begin by arguing that complex suppression mechanisms are essential if selection between groups is to be strong enough to establish complex cooperation. Then they must show how these complex suppression mechanisms can be established when no suitable suppression mechanisms are already in place.

In contrast, Evolutionary Management Theory depends on individuals acting only in their own immediate evolutionary interests. It demonstrates that this is sufficient to explain the evolutionary emergence of complex cooperative societies that are organised by appropriate governance/management.

The essence of the epiphanous realization that I had in my family room was that a system of global governance that manages a planetary civilization is functionally equivalent at an abstract level to the genetic apparatus that manages cells. I saw that a system of global governance had the potential to manage nation-states and other members of a global civilization in such a way that their interests were aligned with the interests of the civilization as a whole. Global governance would have the power to support cooperation and to punish free-riding and theft, particularly as it manifested as war between nation-states.

This would be a repetition at a global scale of the process which produced the United States of America. Various States in North America that were previously independent were united by a system of Federal governance. This system of governance had the power to, for example, disarm states and make further war between them unthinkable. Both the United States of America and a unified, cooperative global society would have the same systemic architecture identified by Evolutionary Management Theory.

In my epiphany, I also saw that the emergence of life in the form of the first proto-cells was made possible by the same kind of systemic architecture. It is generally accepted within evolutionary science that RNA-like molecules emerged in association with early life before DNA emerged. Evolutionary Management Theory envisages that RNA-like molecules and associated processes fulfilled the role of management in the emergence of the first simple cells. They developed the capacity to manage proto-metabolisms.

Proto-metabolisms probably first arose as self-producing, autocatalytic sets of peptides (peptides are smaller versions of proteins). These proto-metabolisms reproduced through a process of collective autocatalysis. In this process, the formation of every member of the autocatalytic set of molecules is catalysed by at least one other member of the set. Consequently, the formation of every member of the set, and therefore the formation of the set as a whole, is boosted by catalysis.

Such a set is self-producing and capable of persisting indefinitely given favourable conditions. However, a self-producing autocatalytic set does not qualify as life in a meaningful sense. This is because its ability to evolve complex functionality is extremely limited.

The limited evolvability of autocatalytic sets is a consequence of the familiar cooperation barrier that impedes the emergence of complex cooperation at all levels of organisation of living processes. The existence of this limited evolvability is easily seen: imagine a peptide that, if it were a member of the set, would significantly improve the ability of the set to grow and persist. It might do this by catalysing a particular reaction that benefits the set as a whole. However, this cooperator peptide will not become a member of the set if its formation does not happen to be catalysed by an existing member of the set. There is no guarantee at all that this will be the case. It does not matter how beneficial the cooperator peptide would be for the survivability of the set as a whole. If it is not catalysed, it will not persist in the set.

Furthermore, imagine a free-rider peptide whose formation is catalysed by members of the set, but which does not contribute anything to the set in return. For example, it might not catalyse the formation of any other member of the set. Instead, it may catalyse other side reactions that drain resources from the set, including resources that may be necessary for the formation of cooperator peptides. As a result, such free-riding peptides will tend to undermine the continued survivability of the set and the cooperative relationships that constitute it.

RNA-like molecules possess the potential to enable an autocatalytic set to overcome this cooperation barrier. This is because RNA-like molecules themselves often have catalytic capacities. For example, they may have the ability to enhance the survivability of an autocatalytic set by catalysing the formation of cooperator peptides that can contribute positively to the set, but whose formation is not otherwise catalysed by the set itself. Furthermore, RNA-like molecules may have the ability to catalyse processes within the set that prevent free-riding peptides from persisting and draining resources from the set.

RNA-like molecules have the potential to manage an autocatalytic set in this way because they are larger in scale, more stable and less reactive than the peptides they manage, and they can function and reproduce independently of the dynamic processes that reproduce the set. Often, this dynamical separation is achieved because the processes that constitute and reproduce management are larger in scale and unfold significantly more slowly than the processes that constitute and reproduce the set. The RNA-like molecules do not become just other members of the set, at the same level of organisation as the original autocatalytic set.

These capacities give RNA-like management the power that living processes must possess if they are to be capable of managing an

141

organisation of entities: They must be able to stand outside the organisation's dynamic processes, act across them, and influence them without being influenced in return.

This is the essence of power at any level: the ability to influence without being influenced in return. For example, if a king can be harmed easily by any of his subjects, he will not have the capacity to rule over them and manage them. If he has the ability to determine whether his subjects live or die, but they also have the same capacity to terminate his life, he will not have the power to govern his kingdom. He will not have control over his subjects. He will likely be a very temporary king. Such a king will be like a CEO whose employees each have the ability to hire and fire him. The CEO will not have the power to manage employees so that they work in the interests of the corporation.

These insights into the potential of processes at one level of organisation to constrain and manipulate a lower level were prompted by the 1985 book, Evolving Hierarchical Systems, written by evolutionary scientist Stanley Salthe. When I meet evolutionary scientists at conferences, I often mention this book and suggest that it will eventually be seen as one of the top five scientific books of the 20th century. If they agree, as some do (but very few), I know immediately that I am talking to someone who has at least some capacity for what I now refer to as metasystemic cognition. Someone worth taking seriously.

Broadly then, evolvable RNA-like molecular processes have the potential to manage a self-producing autocatalytic set of peptides, proteins, and other molecules in the same way that a ruler or government manages a human society. This management has the potential to enable complex cooperation to emerge and evolve within the proto-metabolism.

However, this brings us again to the questions that arose regarding human managers/governors: Why would the genetic apparatus use its power in this way? Can the emergence of such management be explained by evolutionary processes accepted and known by mainstream evolutionary science?

Again, at a high level of abstraction, the answers to these questions are similar to those that apply at the human level. Initially, RNA-like molecules are likely to have predated on autocatalytic sets of peptides and other molecules. Due to their collectively-autocatalytic organisation, these sets would have been a rich source of smaller-scale molecules that could, for example, be used by the RNA-like molecules as components for their own self-production. The RNA-like molecules could use their catalytic power to plunder resources from an autocatalytic set, and then drift on to repeat the process with other sets that they encountered.

Eventually, RNA-like molecules could emerge that would remain with an auto-catalytic set and harvest an ongoing stream of benefits from it. Such an autocatalytic set would become a proto-metabolism in relation to the RNA-like molecules. It would be a short step from this to the evolution of RNA-like molecules that could use their power to manage a proto-metabolism in ways that enhance its productivity, thereby enabling the manager to harvest a greater stream of benefits from the set. These RNA-like managers farmed autocatalytic sets. As we have seen, due to the seriously limited evolvability of unmanaged autocatalytic sets, management had the potential to manage them in ways that significantly enhanced the harvestable benefits.

Self-producing autocatalytic sets crossed the threshold to life only once the limitations that restricted their evolvability were overcome by appropriate management. It was only when this occurred that these self-producing organisations developed the evolvability that characterises living processes, and could thereafter adapt and evolve as coherent wholes.

It is worth emphasising the significance of this transition. Before the transition, the only catalysts that could be incorporated into a set were those whose formation happened by chance to be catalysed by existing members of the set. This seriously limited the possibility space of adaptations that could be explored by such a system. It could not discover and instantiate any of the numerous potential adaptations that depended on catalysts that were not themselves catalysed. This was the case no matter how big a contribution could be made by such an adaptation to the survivability of the set. Similarly, a set would be unable to incorporate processes that could inhibit free-riding side reactions that would otherwise undermine the survivability of the set,

143

except where those inhibitory processes happened fortuitously to be catalysed within the existing set.

After the transition, management by RNA had the potential to enable the incorporation of a much wider range of catalysts into a set, and to support processes that inhibited free-riding. This greatly increased the space of adaptive organisational possibilities that could be explored. The evolvability of proto-metabolisms was therefore enhanced significantly by the emergence of effective management. Such an organisation had a much greater potential to discover cooperative relationships that enabled the organisation to adapt in ways that could improve survivability. This enabled self-producing molecular organisations to emerge that could adapt and evolve as coherent wholes. This in turn enabled managed sets to differentiate internally and to incorporate adaptive divisions of labour and specialisation, enabling complex functionality to emerge. Internal cooperative differentiation of this kind is seen in all major cooperative evolutionary transitions once the relevant cooperation barrier is overcome comprehensively.

For all these reasons, life could not properly be said to have emerged until self-producing molecular organisations were taken over by the systemic architecture that enabled the cooperation barrier to be overcome at the molecular level.

Once I had thought through in detail the ideas that first occurred to me as a flash of insight, it seemed obvious that there was an abstract functional equivalence between the systemic architecture that emerged at the origins of life, and the global management that can create a cooperative global society. This obviousness was reinforced by the fact that similar dynamical architectures can also explain the emergence of the other major cooperative evolutionary transitions.

However, it was a long time before I realized that this similarity is only obvious in retrospect, i.e., only once one has 'seen' the similarity by building mental models that demonstrate its existence. Then, but only then, is it obvious. It will not be obvious to you until you build the requisite mental models.

This transition from unintelligibility to obviousness has many examples in the history of science. To experience a striking example, put yourself in the frame of mind of a person living before any general understanding of gravity had been developed. You would not see any relationship between what is observed in the heavens (with the sun, moon, planets, and stars moving about endlessly), and what happens here on Earth when we let go of an object and it falls to the ground, ceasing to move further. The two classes of phenomena, those in the heavens and those on Earth, would seem to you to be totally unrelated.

This was the situation that was faced by Galileo back in the sixteenth century. He was undertaking experiments to try to understand the trajectory of a projectile that is launched on the surface of the Earth. He discovered that the projectile would follow a path that could be described precisely. He observed that the projectile moved forward at the same time as gravity attracted it towards the ground. Eventually, the projectile would fall to earth.

Then Galileo saw an abstract equivalence: as projectiles are launched with greater and greater force, they will move further and further before they hit the ground. Then he imagined that a projectile could be launched with such force that its trajectory would encircle the globe, never falling to Earth. He realized that this was what was happening with the apparently perpetual movement of heavenly bodies. The projectile launched with sufficient force would be just like the moon, orbiting the Earth endlessly. He saw that both sets of circumstances could be explained by one, unifying theory. As above, so below. We now take this as obvious, and teach it to school children. Once you see it, it is obvious. It is difficult to un-see it.

However, a major challenge remained if I was to develop a general, unified theory that identifies the mechanisms that produced all the major evolutionary transitions that characterize the trajectory of the evolution of life on Earth. There are a number of important transitions that do not appear to have been organised by the systemic architecture that I had identified. These cooperative transitions could not be explained by the emergence of management imposed by external sources of power. The three key examples are the transition to multicellular organisms from single cells, to insect societies from individual insects, and to human tribes from individual humans.

It took many months for me to build mental models that enabled me to understand fully how these transitions were achieved. For my cognitive development, this effort proved very beneficial. Overcoming the challenge demanded further recursive testing and improvement of my model-building capacities.

For example, I had to build mental models of the three transitions I have mentioned, and inhabit those models for hours at a time as I modified and refined them. This involved building dynamical models in my mind; seeing them from different perspectives; zooming in to scrutinize more closely the dynamical operation of particular components of the models; modifying key aspects of the models and then using the changed models to track the consequences of those changes; comparing models at higher levels of abstraction to see if there are commonalities between the models at different levels; and so on. I had to do this consciously and intentionally. Further, it was necessary to be aware at a meta-level of the thinking strategies that I was using, and to recursively improve them as I proceeded to think my way through the issues.

To undertake this recursive self-improvement effectively, it was essential that I was able to be aware of my thought processes. I had to be able to see them as object, ideally in real-time. At the time, I was hoping that my meditation practice would help me to enhance this capacity. However, when I first began to develop the complex, dynamical mental models that captured the essential elements of these transitions, my thinking tended to be somewhat simple, vague, and ill-defined. But as I continued to put in time and effort, models that were clearer and more useful began to emerge from the mist and fog.

Eventually, I saw that two forms of management existed that had the capacity to enable complex cooperation to emerge. As I have discussed in detail, the first form was where management was external to the organisation being managed. The second was where management was internal to the entities being managed and distributed across them. Of course, both architectures might be operative in any particular transition.

146

Again, the germ for this realization came from Stanley Salthe's theory of evolving hierarchical systems. He argues that entities can be controlled not just by external processes, but also by their internal constituents. These can 'hard-wire' the entities with particular characteristics. These internal constraints can hard-wire an entity in the sense that they can control its behaviour but are not changed during its life. In other words, the hard-wired internal constraints can influence the behaviour of an entity, but without being influenced in return.

Examples of hard-wired internal constraints include inherited genetic predispositions and inculcated cultural predispositions. It is easy to see that genetic predispositions embodied in the genetic apparatus can control the behaviour of individual cells within a multicellular organism, individual insects within an insect colony, and individual humans in some circumstances. And it is easy to see that inculcated cultural predispositions such as norms and morals can control the behaviours of individual humans, provided the predispositions are internalised sufficiently.

However, it is much more difficult to see how these predispositions that are internal to individuals could somehow combine to control a group of organisms. How could they control a group in the way that an external manager can? And how could they capture the benefits they create when they support complex cooperation within a group?

We will see that internal predispositions can meet both these requirements if a particular condition is met. This condition is that the same set of predispositions, whether genetic or cultural, are installed in each and every member of the group. When this condition is met, a predisposition that generates cooperation will capture all the benefits created by this cooperation. This is the case even though a particular cooperator who uses its resources to benefit others may end up worse off than average. However, the others within the group that benefit from this cooperation will also contain the set of cooperative predispositions. Consequently, the cooperative predispositions will capture all the net benefits produced by the cooperation it generates.

But this just seems to lead to a bigger problem: What plausible process would ensure that the relevant set of predispositions is reproduced in each and every member of the group? This problem is particularly significant given that if any free-rider or other non-cooperator emerges in the group, it will receive benefits from cooperators without contributing to the group. As a result, it will tend to outcompete the other members, and its progeny will potentially take over the group.

A closer examination of each of the three examples will reveal how this condition is met. In sexually reproducing multicellular organisms, each individual organism originates as a single cell. Therefore, all the cells in such an individual tend to contain the same genes. In particular, they will all contain any set of genes that predisposes the cells to cooperate.

In complex multicellular organisms, this set of genes will also contain genes that predispose cells to produce an immune system. Amongst other things, the immune system will tend to destroy mutant cells that escape the control of the original set of genetic predispositions. In humans, if some of these mutant cells escape the immune system, their competitive advantage can result in rapid proliferation, manifesting as cancer. However, in the great majority of cases, these kinds of mechanisms ensure that each and every cell within a given multicellular organism contains the same set of genes, and is controlled by them.

Due to a similar process, the individual members of insect societies share the same set of genetic predispositions to a significant extent. This is because they are closely related. Insect societies also often include immune system-like mechanisms that cleanse the society of any free-riders that emerge.

The members of human tribes are often not related closely. As such, there is no likelihood that they will each include the same genetic predispositions. However, they share an enculturation process that tends to inculcate all members of the tribe with a shared set of cultural norms and beliefs that shape their behaviours and interactions. These cultural predispositions are generally entrenched in all individuals born into the tribe, including through socialisation. This entrenchment is typically deepened further by a shared system of supernatural beliefs and associated group rituals. As a result, the members of the tribe tend to carry the same set of cultural predispositions.

However, because a cooperative group is susceptible to being undermined by any free rider that arises, tribes need to have a powerful mechanism that ensures that all members exhibit the relevant predispositions, and that any free riders that arise are quickly punished. Typically, this was achieved in tribes by the inclusion of cultural predispositions to punish or expel any individuals that behaved as if they did not contain the particular set of cultural predispositions that characterised the tribe. This kind of mechanism was generally orchestrated by cultural predispositions that organised the continual surveillance of all members of the tribe by all others. This included gossip as a means for sharing information about possible transgressions. Such a mechanism also tended to police and enforce the possession of pro-social genetic predispositions.

Persistent breaching of tribal norms would generally result in punishment that was administered by the tribe as a group and that was orchestrated by cultural predispositions. This could include exclusion from the tribe. In effect, expulsion would generally amount to a death sentence. Typically, individuals isolated from a tribe did not survive. For any members of a tribe who persistently breached the beliefs, norms, and morals of the tribe, life would tend to become solitary, poor, nasty, brutish, and short. This was the case irrespective of whether transgressors had any rational basis for their failure to act according to group beliefs and norms.

I use the term Distributed Lower-Level Management (DLLM) to refer to this form of management and to distinguish it from External Management (EM).

Of course, many members of modern human societies adhere strongly and unquestioningly to various sets of religious and cultural beliefs. In many instances, these cultural predispositions tend to cause individuals to act in pro-social ways. Often, this pro-social internal management complements any prosocial management established by governments or other EM.

However, DLLM by itself is unable to organise large-scale cooperative human societies. This is due to the practical difficulties it encounters as the scale of societies increases. For example, it is impossible to ensure that all members of a large-scale society are inculcated successfully with the relevant set of cultural predispositions and that transgressors are eliminated quickly.

Nevertheless, 'world religions' that treat all ethnicities equally have played a significant role in the successful formation of large-scale empires and multi-cultural societies. The sets of cultural predispositions that are spread by these religions made the external management of these societies easier and more effective. Modern societies in which many citizens adhere to a pro-social religion are less challenging to manage. As a result, the systems of religious beliefs that have survived and spread over the past 10,000 years tend to be those that have facilitated the emergence of successful external management.

However, the relative invisibility of DLLM often leads to its effects being misunderstood. For example, human tribes have often been held up as examples of how cooperative human societies can be achieved without the existence of centralised power. They are not organised by a dominant ruler or manager. On the surface, human tribes seem to have achieved harmony and order through spontaneous bottom-up processes, rather than by hierarchical top-down power. The egalitarian nature of tribal societies gives hope to those who are attracted to the idea that human societies can be organised successfully without the need for external governance.

Many seem to believe that when such a tribe encountered a challenge, the members would meet around the campfire as equals and freely discuss the options open to the tribe. A decision would be arrived at without force or coercion. Reason, not power, would prevail. However, once an individual has developed mental models that reveal the otherwise invisible mechanisms that orchestrate the actions of tribes, it becomes clear that this is not how tribes are organised.

Like large-scale hierarchical societies, tribes can only achieve complex cooperation if a source of power exists that can support and enable beneficial cooperation. Importantly, this source of power also must be able to punish behaviours that would undermine cooperation, and, if necessary, eliminate them. DLLM often achieved this by organising group punishment, and by predisposing members of the tribe to feel moral outrage towards the transgressors.

As a result, life for idiosyncratic free thinkers in a tribal society could be even more unpleasant than if they lived in a small country town in Australia or in a town of similar size in the center of the United States. They would be continually under surveillance, and their behaviour would be judged continually against standards that often could not withstand intelligent scrutiny.

Furthermore, the DLLM mechanism tends to be invisible to members of groups that are organised by DLLM. The members will have been inculcated with the relevant set of cultural predispositions. However, they will generally be completely unaware of the role that their beliefs and norms play as part of a system of DLLM. Often, they will believe that they have freely chosen their religious beliefs and moral principles.

For example, take the case of conservative Christians born and raised in evangelical Christian families in the mid-west of the United States. It would be extremely unlikely for them to realize that, if they had instead been raised in a family of devout Hindus in a farming community in central India, it would be extraordinarily unlikely that they would be a committed evangelical Christian.

*   *   *

I should mention here that the impact of my epiphanous experience was not just intellectual. In combination with the theories I developed subsequently, it seems eventually to have also produced a kind of emotional and motivational epiphany. This overwhelmed me one day when I was in the process of opening the front door of our house to go to work. For some reason, the plight of the millions of children who die each year of malnutrition came into my mind, including the many millions more whose intellectual development is stunted due to inadequate access to food. It hit me that each and every human on the planet has a responsibility to stop this from happening, particularly given that the world can easily produce enough food to prevent this.

I burst into tears as I realized that my theories identified the only way in which these and other horrors, such as war, could be overcome permanently. Humanity needs to take the next great step in the evolution of life on this planet by instituting a cooperative and sustainable global society underpinned by an appropriate system of global governance. I saw that my evolutionary insights came with a very heavy set of responsibilities. I had to do whatever I could to make this transition happen.

In an instant, I transformed permanently from a person who was motivated primarily by ego to develop and publish his ideas, to a person who was primarily motivated to serve goals and purposes much larger than himself. This magnified the strength of my motivation manyfold. If this emotional epiphany had not occurred, I would not have produced what I have. It is still the force that moves me today in writing this book.

13.

Publishing my Theories

By early 1993, I had reached the point where I could begin to write down my theories and submit papers for publication in international, peer-reviewed science journals.

Rather than start with a paper explaining the complex insights I have just outlined, I decided to begin with something less ambitious. I commenced with a simple paper that focused on a limited but important problem in evolutionary science: how is it that sexual reproduction tends to out-compete asexual reproduction in multicellular organisms?[7]

This is seen as an issue because, on the surface, natural selection would appear to favour asexual reproduction. An organism that reproduces asexually (i.e., by producing clones of itself), would appear to produce twice as many reproducing offspring as would any member of the same species that reproduces sexually. This is because half of the progeny of the members that reproduce sexually will be males who do not reproduce themselves directly. All other things being equal, all the progeny of the asexual members will reproduce. Consequently, asexual members will rapidly increase relative to sexual members, and take over the population.

This issue seems to be far divorced from my ideas about the mechanisms that drive the trajectory of evolution toward complex cooperatives of greater and greater scale. However, they both arise from a particular perspective that generated much of my evolutionary thinking. This perspective views evolution as resulting from the operation of mechanisms that are 'intelligence-like'. These mechanisms

---

[7] Stewart (1993) – see References

can be viewed usefully as 'intelligence-like' in the sense that the mechanisms solve adaptive challenges.

Another way of putting this is that the mechanisms that discover evolutionary adaptations do so by searching a 'possibility space' that encompasses all possible adaptations. The mechanisms search this 'possibility space' for adaptations that enable organisms to adapt to the environmental challenges that they face.

The most familiar example of such an 'intelligence-like' mechanism is adaptation by natural selection. In its most simple form, this evolutionary mechanism searches possibility space by producing mutations. Each mutation is a possible adaptation. Natural selection then evaluates each possible adaptation to assess whether it solves some adaptive challenge. Where it does, the mutation tends to outcompete other possibilities and eventually take over the population.

A mechanism that searches possibility space by randomly producing mutations is a very ineffective method for discovering adaptations. Blind trial-and-error is a very slow, inefficient, and wasteful way to discover adaptive advantages. In the sense I am using here, it is not very intelligent.

A mechanism would be less wasteful if it could search possibility space using a method that tested only genetic possibilities that were more likely to be adaptively successful. Instead of trialling random possibilities, it could try out only possibilities that had a greater probability of being advantageous. A mechanism that could do this would be able to outcompete alternative mechanisms that trialled possibilities randomly.

Sexual reproduction is such a mechanism. In addition to searching possibility space by trying out random mutations, it also searches by trying out different combinations of existing genes. Existing genes have already demonstrated a capacity to contribute to the adaptation of the organism to its environment. Consequently, a different combination of these existing genes might have a higher probability of success than a random change in any of the genes themselves.

Sexual reproduction combines the genes of each parent and shuffles them in a process known appropriately as recombination. This can create completely novel combinations of genes. Sexual reproduction is smarter than asexual reproduction. It is less wasteful because the method it uses to explore possibility space is more likely than random mutation to produce better-adapted offspring.

My first paper argued that when the environment changes, sexual reproduction is able to outcompete asexual reproduction. This is because it is better at adapting the organism to new circumstances. This was not controversial or novel. But the paper went further. It demonstrated that even if a species was not encountering an environment that was continually changing, sexual reproduction could nevertheless outcompete an emerging clone. This is because the initial growth and success of the clone itself would impact the environment faced by the organism. These environmental changes themselves would provide an adaptive advantage to the members of the population that reproduce sexually.

From a broader, 'big picture' perspective, the key idea here is that evolution itself evolves. Sexual reproduction emerged and spread because it was better at adapting organisms. The mechanisms that search for and produce evolutionary adaptation tend to be improved themselves as evolution proceeds. Furthermore, the improved mechanisms can then improve themselves, and so on, indefinitely. Evolvability, the ability to discover and implement better adaptations, has tended to improve recursively in this way throughout the evolution of life on Earth. This is a central aspect of the trajectory of evolution.

As we have seen, before the genetic apparatus emerged, the evolvability of autocatalytic sets, the first self-producing organisations of molecular processes, was very limited. The emergence of RNA-like managers massively enhanced their evolvability. Evolvability was further enhanced by the emergence of managers who could explore possibility space digitally, i.e., by random changes in a genetic code. This greatly expanded the possibility space that could be explored, and enabled it to be explored more systematically.

As we have also seen, sexual reproduction represented an additional major transition in the ability of the evolutionary process to discover and implement effective adaptations. However, until

155

evolvability improved further again, evolutionary mechanisms operated by trying out possibilities through reproduction. Organisms did not try out heritable changes during their life. They could adapt somewhat during their life, but no matter how successful the changes, they could not be passed on to their offspring or to other members of the population. Any superior adaptations that were discovered by an organism during its life died with it.

This changed with the emergence of cultural inheritance. However, only in humans has it developed in a complex form. Most multicellular organisms can adapt during their lives somewhat by learning. For example, through the process known as operant learning, they can try out different behaviours as they interact with their environment. Any that are positively reinforced in particular circumstances will tend to be repeated when similar circumstances are encountered again. Through this process, the organism learns what works in the world. They learn by consequences, as did gene-based evolution before it. Through trial-and-error, the organism searches for and discovers behaviour that is adaptive.

The emergence of the cultural evolutionary mechanism converted these adaptive discoveries into evolutionary adaptations that could be passed on to other members of the species. No longer did they disappear when the individual died, like tears in rain. Learned adaptive discoveries could now accumulate as an evolving cultural inheritance.

A further significant increase in evolvability occurred when trial-and-error experimentation was able to be undertaken in an organism's mind, rather than in the world. This was achieved by the development of an ability to build mental models that could be used to identify effective adaptions.

This capacity enabled organisms to build mental models of their interactions with their environment. The models could then be used to simulate the outcome of particular behaviours. Evolvability was enhanced significantly once organisms could use their mental models in this way to predict the outcome of behaviours that they had never tried out in the world previously. This enabled them to evaluate the adaptive effectiveness of novel behaviours 'in their heads'.

For example, early humans could build a mental model of how they might construct a shelter in which their family could sleep safely at night. They could use this model to simulate how they might use different materials to build an effective shelter in places where the materials they had used previously were not available. They did not have to experiment in the real world with how they might use different materials. Instead, they could modify the materials in their mental model, and simulate the outcome of doing so.

Evaluating alternatives in the virtual reality of one's mind is far more efficient and often safer than doing it in the real world. And the discoveries that are made through the use of mental models can be transmitted to others through various mechanisms. It is possible to identify various steps in the evolution of transmission mechanisms: e.g., from imitation; to demonstration; to oral language; to recorded symbols such as writing; to printing; to film; to the internet; and so on.

The capacity to construct mental models can also be seen to have moved through different stages: e.g., from models that comprise representations of concrete phenomena; to models that can also include abstractions; to models that can adequately represent complex phenomena; and so on.

Furthermore, external aids such as paper, calculators, computers, artificial intelligence, and so on have enhanced the evolvability of these mechanisms.

Finally, each new kind of entity that emerges from a cooperative of smaller-scale entities during the evolution of life can be considered from the perspective of evolvability. Each such larger-scale entity is constituted by processes that enable it to evolve, and the evolvability of these processes also evolves. At the human level, it is possible to consider the evolution of the evolvability of collectives such as corporations, nation-states, and if it emerges, a global civilization.

It is worth noting here that this capacity to build and operate mental models requires consciousness. Broadly, this is because consciousness emerges when there is a sub-system that comprises internal representations (the 'object') that are managed and manipulated by a 'subject'. Together, these constitute a 'subject-object subsystem.'

157

The capacity to use and manipulate mental models comprises such a subsystem. Consciousness arises for the subject in such a subsystem because there is something it is like to be a subject that manages and manipulates mental models. But it was nearly twenty years before I developed an effective and plausible model of consciousness that provided this understanding.[8]

The acquisition of a conscious ability to construct and use mental models provided organisms with the capacity to look ahead and anticipate the future consequences of their possible actions. This enabled them to evaluate possible adaptations not just based on how they would work in the here and now. Instead, they could now adapt in the present in ways that took into account their predictions about future events.

A further major transition in evolvability occurs when organisms develop the ability to construct mental models of the evolutionary processes that will shape their future evolution. They will be able to use these models to identify how they need to adapt and evolve in the present to ensure that they will survive and thrive indefinitely into the future.

For example, they will be able to identify the trajectory of evolution, locate themselves along the trajectory, and see what they will need to do to remain aligned with it in the future. In particular, they will be able to identify how they will need to organise themselves socially and enhance their evolvability to avoid being selected out of existence.

The trajectory of evolution identifies how organisms need to be adapted if they are to avoid becoming casualties of selection. Discovering the direction of evolution will therefore enable them to see how they will need to live their lives and adopt values and goals that will enable them to survive and flourish indefinitely into the future. It will enable them to see what they need to do to contribute successfully to the future evolution of life in the universe.

Humans are currently just beginning to enter this next great transition in evolvability.

---

[8] Stewart (2022) – see References for full citation

But already, evolvability has increased enormously since life first emerged on Earth. It took gene-based evolutionary processes many millions of years of trial-and-error to discover how to produce heavier-than-air flight. Humans achieved a form of such flight in only a few centuries of technological innovation. Technological development was enabled by culture-based evolutionary processes and by intelligence that is powered by conscious model building.

Why did I have this propensity to view the evolutionary mechanisms that evolved living processes as having themselves evolved? Why was I somehow predisposed to attempt to develop a general theory of evolvability and its role in the trajectory of evolution? Looking back, it was no accident.

As I have outlined, at a young age I stumbled upon the possibility that I could consciously improve my intelligence and other abilities. Inspired to some extent by books I had read in my early teens, I realized that I could intentionally work on myself to enhance my capacities. I did not see my adaptive capabilities as being fixed or given. I was, and always would be, a work in progress. In the terms used by Gurdjieff, I saw the possibility of becoming a self-evolving being. Propelled mainly by chance events, I had begun in my teens to put some of these ideas into practice. I developed an ability to recursively enhance my own problem-solving capacities. And so on.

Consequently, when I began to think about evolutionary mechanisms like gene-based natural selection, I was predisposed to view them as processes that searched for and discovered solutions to adaptive challenges. Furthermore, I did not see them as fixed and given. I was predisposed to ask in relation to evolutionary mechanisms the kinds of meta-questions that I also asked myself about my own problem-solving abilities. In relation to my own thought processes, I asked questions like: are my existing processes limited in particular ways? How can these limitations be overcome? How can I amend my thought processes to enhance their capacity to solve challenging problems?

In relation to developing a general theory of evolvability, this led me to ask the following kinds of questions: what processes does a particular evolutionary mechanism use to discover adaptive improvements? What were the evolutionary mechanisms that preceded it? How does the evolvability of this mechanism improve on these previous mechanisms? In what ways is the particular mechanism still limited in its ability to discover innovative adaptations? What specific changes to the mechanism could overcome these limitations? Did such an improved mechanism actually emerge in evolutionary history? Why? Once those improvements were made, what limitations would remain? How could those remaining limitations be overcome? Is there a trend in this sequence of enhancements? Are the mechanisms at the human level the result of a trend that began at the gene-based level? Is it possible to use the answers to these kinds of questions to identify a trajectory of the evolution of evolutionary mechanisms? Can this lead to a unified theory of evolvability that is applicable across all levels of organisations? And so on.

*   *   *

My first paper, 'The Maintenance of Sex' was published in the journal Evolutionary Theory in 1993. The second, 'Metaevolution', was published in the Journal of Social and Evolutionary Systems in 1995.[9] Metaevolution presented the 'big picture' theory that I have outlined above. It argued that evolution embodies a trajectory towards increasing integration and cooperation, and identified the mechanisms that produce this trajectory.

I coined the term 'Metaevolution' to refer to the evolution of evolution. I thought that this focus on the evolution of evolution was apt because I saw that each step in the evolutionary trajectory towards increasing cooperation also represented an increase in evolvability.

I was propelled in this direction because, as I have indicated, I was predisposed to view evolution through the lens of evolvability. This led me to see that initially, when a new level of living organisation was beginning to emerge during evolution, the entities at that level tended to compete against each other. This competition impeded the emergence of

---

[9] Stewart (1995) – see References for full citation

complex cooperation amongst the entities, constituting what I have referred to as a cooperation barrier.

The barrier prevented the evolutionary process from fully exploring the adaptive benefits of cooperation between entities. Cooperative divisions of labour, specialization, and other synergies could not be discovered and implemented, no matter how beneficial they might have been adaptively. Because complex cooperation could not persist in a stable fashion in the population, it could not be selected into existence.

In other words, the cooperation barrier prevented the relevant evolutionary mechanism from fully exploring the space of adaptive possibilities that involve complex cooperation. As such, the barrier represented a serious limitation in evolvability. From this perspective, my Metaevolution paper was about how the emergence of appropriate management could enhance evolvability by overcoming the limitations imposed by the cooperation barrier.

At the time, I thought that this was a completely novel, insightful, and useful perspective. However, I decided eventually to drop this meta-evolutionary view of the evolution of cooperation. As I gained more experience in mainstream evolutionary science, I realized that I should keep my papers as simple and straightforward as possible. Instead of trying to pack each paper with radical new ideas, I should focus on one, or at most two, at a time. I decided to dumb down my papers and choose simplicity and clarity over complexity and novelty.

This strategy has been productive. I have yet to write a paper that has not been published. But the road has often been rocky. About half of my papers have been peer-reviewed and accepted by the first journal I submitted them to. But the other half have been rejected at least once before they were eventually published.

I had three more papers published in 1997.[10] Two in the Journal of Social and Evolutionary Systems, and the third in the journal Artificial Life. The first was about the evolution of genetic evolvability. The second was an attempt to get artificial life practitioners to set up

---

[10] Stewart (1997a), (1997b) and (1997c) – see References for detail

artificial life systems in such a way that they would undergo major cooperative evolutionary transitions. The third paper discussed whether the trajectory of evolution was progressive in evolutionary terms.

14.

Working for Evolution

By the end of 1998, I decided it was time to write a book that put together my ideas about the trajectory of evolution. A major focus of the book was to be the development of an evolutionary theory that covered both human and biological evolution. The book would present a theory that was unified in the sense that it identified general principles that apply to the evolution of all living processes, not just to a particular subset of them.

Of course, such a unification could not be achieved only at the surface level. Details differ substantially across the enormous variety of living processes that have emerged on this planet. Instead, a unified theory would need to consider aspects of evolutionary phenomena that come into focus only at higher levels of abstraction.

Such an approach had the potential to identify what is common across all specific instances. It could show that the profusion of details that come into focus at lower levels of abstraction do not contradict the patterns that are evident at higher levels. Without such a model-building strategy, abstract and unifying patterns would likely be obscured forever by the plethora of detail. This kind of strategy is common to most successful research programs in science.

But the main reason for writing a book was that I was discovering that such a unified, 'big picture' evolutionary theory had significant implications for humanity, here and now—for the way we live our lives and organise ourselves socially. These findings were not suited for publication in science journals. There were several reasons for this.

First, the discoveries that had implications for humanity tended to cross many disciplinary boundaries. In particular, I was finding that my evolutionary theorizing was leading me into fields as diverse as economics, ethical philosophy, political science, and anthropology and sociology. As far as I knew, there were no reputable journals that would publish articles of sufficient diversity and length to outline the results of my thinking.

Second, science journals would tend to view my theory as highly speculative and not rigorous enough to qualify as hard science. Given the rules and methodologies used at the time to define proper science, such an assessment would be reasonably accurate. Any comprehensive evolutionary theory that set out to explain and make predictions about complex aspects of human evolution was likely to run afoul of the standards of mainstream science.

If such an evolutionary theory were to be comprehensive, it would have to encompass the evolution of human political, economic, and social systems. However, these domains are generally studied as part of the humanities, not by mainstream science.

The methods and research strategies used by mainstream science have failed miserably to make any significant progress in the humanities. Where the humanities are forced to use these science-based methods, such as in much of American academic psychology, it tends to produce findings that are rigorous and evidence-based, but that are trivial, uninteresting, and largely irrelevant. Generations of students who enroll in University courses in psychology have been seriously disappointed. Instead of learning something insightful and useful about their own psychological functioning, all they get in the main is 'rats and stats'.

As I have mentioned previously and will discuss in greater detail later, mainstream science is ineffective at understanding complex, dynamic phenomena. Unfortunately, most aspects of reality that are of interest to humans are comprised of dynamical systems of this kind.

As we shall see, a new kind of science and a new level of cognition are necessary for science to make worthwhile progress in these areas.

I was not prepared to restrict my work to what was considered acceptable to mainstream science. The intuitions that had driven my thinking since I was a teenager were that a comprehensive evolutionary theory had the power to answer key questions about human existence: Where do we come from? What are we? Where are we going to? and What should we do with our lives? I was finding increasingly that these intuitions were accurate. I was making real progress in using an evolutionary worldview to address these fundamental issues.

I had never been interested in studying evolution from an external perspective. Current mainstream science tends to study its subject matter in this way, from the outside. It tends to proceed as if scientists are separated from reality by a pane of glass. Science is at great pains to exclude scientists and their subjectivity from their study of phenomena.

However, this separation is shattered once the phenomena being studied include scientists, their ethics, any meaning and purpose that they find in their lives, as well as the societies, economic, and political systems in which they are embedded, and how all these phenomena are likely to coevolve into the future.

As I was discovering, the more my evolutionary theorizing progressed, the more that the discoveries I was making tended to affect my own goals and the way that I viewed myself and my being in the world. I, my actions, and my strategies were all relevant variables in the theories that I was developing.

This was the reality that confronted me at the end of 1998. If I wanted to write about the implications of a general theory of evolution for the human condition, I would have to write a book. Reflecting this, the title of the book was to be 'Evolution's Arrow: The Direction of Evolution and the Future of Humanity.'

Unfortunately, I would not be able to write the book at work. Long hours of uninterrupted thinking and writing were required. Work was great for writing and editing papers, but writing a book from scratch was impossible. Fortunately, Australian Government employees were entitled to long service leave of 13 weeks after 10 years' service, and pro rata for service after that. This could be taken on half pay for double the period of leave. Combining this with my accumulated annual leave, I calculated that I could afford to have up to a year's leave on half pay. My wife agreed to the proposal.

In the 25 or so years since then, I have continued to develop and refine both my evolutionary theory and its implications for myself and for humanity more generally. When I get near the end of this part of the book that outlines my developmental odyssey, I will summarize the main findings and implications of my evolutionary theorizing.

For those interested in exploring particular issues in greater depth, I will also provide references to papers of mine that are relevant to each issue. But here, I will confine myself to giving an example of one of the key issues that preoccupied my thinking at the time. It demonstrates the powerful potential of the evolutionary worldview and metasystemic cognition to answer the big existential questions that face us all.

I will explore how an evolutionary worldview can fundamentally change how we decide how to act and behave. It can transform how we see ourselves and produce a new kind of human being.

I will begin by taking as an example the role of ethical and moral principles in shaping human behaviour. I will demonstrate that the role of ethical and moral principles will diminish in importance as our cognitive capacities increase as we evolve into the future. Ethics, norms of behaviour, and morals will increasingly become redundant.

We have discussed earlier how norms and morals played a critically important role in the evolution of cooperative human tribes and societies. Entrenched by socialisation and often also by religious systems and genetic predispositions, norms and morals were part of the system of management that was hard-wired into members of the society. These predispositions caused members to behave in ways that enabled complex cooperation to be sustained within the group. Selection driven by competition between groups favoured societies that were managed by systems of norms and morals that were better at organising the society effectively.

The members of any given tribal society were not free to choose to change their tribe's norms and morals at will. As discussed earlier, the effectiveness of management depended on it being hard-wired across the members of the society. If the members had the capacity to change their norms and morals at will, it would often have led the tribe to disaster.

This is because the tribal members did not have the cognitive ability to build mental models of alternative systems of norms and morals and use the models to identify the system that was best for themselves and the tribe. A cognitive capacity at the metasystemic level would be necessary for this. The tribe, its organisation, and the effects of changes to its norms and morals constitute phenomena that are highly complex. Even analytical/rational cognition wielded by highly-trained sociologists and anthropologists struggles with this kind of challenge.

Consequently, the evolvability of systems of norms, morals, and associated religious beliefs was very limited. To the extent that they did evolve as circumstances changed, this tended to be driven by costly selection resulting from inter-group competition.

However, social and cultural circumstances have changed significantly since humans lived in tribal societies and since monotheistic religions first emerged. Relevant circumstances continue to change rapidly and are likely to do so even more quickly in the future.

There are obvious disadvantages in continuing to have our actions dictated by inflexible rules and principles established by past evolution. The behaviours that were favoured during our evolutionary history are highly unlikely to continue to lead us to evolutionary success going forward. In fact, it is becoming increasingly likely that they will lead us to environmental destruction within the next 100 years.

To the extent that we continue to maintain our preexisting genetic, cultural, and social predispositions, humanity is likely to be maladapted to the new circumstances. Significant changes to our behaviours are likely to be necessary to ensure we continue to be adapted to changing social and environmental conditions.

However, few humans currently live as if they are aware that the fundamental characteristics that govern their behaviour have been shaped by evolutionary processes. Few view their religious beliefs, values, and ethics as having been fashioned by past evolutionary processes. Many take their religious and ethical beliefs to be true, and see their values as being absolutely appropriate and justified. They tend to be embedded in their emotional states. They fail to see them as evolution's method of motivating particular behaviours that were adaptive in the past environments that shaped them.

Even fewer see their fundamental beliefs, values, and ethics as characteristics that need to be reviewed and evaluated regularly as relevant circumstances change. They do not see them as contingent features that should be adapted intentionally as social and environmental conditions vary.

It is useful at this point to imagine a hypothetical scenario in which life emerges and evolves on some other planet in our galaxy. Imagine that evolution on this planet gets to the point where sentient organisms arise and go through a tribal phase. Consequently, they come to be equipped with entrenched ethical and moral beliefs that organise pro-social behaviours.

Now, consider the possibility that some of these organisms develop higher cognitive capacities. In particular, they become cognitively advanced enough to construct mental models of how their social and physical environment is likely to evolve and change in the future. These models would enable the organisms to assess how they might need to change their goals, values, and ethics if they are to survive the predicted changes. They would be able to predict the selection that would be imposed by changed circumstances, and identify the kind of modifications they would need to make to their behaviour in order to avoid costly selection.

Such an organism would no longer need to depend on selection installing in it the genetic and cultural predispositions that would cause it to behave in adaptive ways. The organism would not have to continue to obey blindly the predispositions that were installed in it by past evolution, including those that may be maladapted to current or future circumstances.

Instead, they would be able to use their mental models to identify how they need to remake themselves intentionally. They would be able to work on themselves consciously in order to develop the capacity to free themselves from the dictates of their genetic, cultural, and social past, including their conditioning. This would enable them to move at right angles to their existing predispositions. They would become self-evolving organisms, able to adapt in whatever ways are demanded as circumstances change. No longer would evolutionary processes such as genetic and cultural selection have to force them to change. They would be able to work it out themselves, and choose to act accordingly.

More specifically, these organisms would no longer have their behaviour guided primarily by ethical and moral principles or rules. When confronted with complex social challenges, for example, they would not have to decide what to do by consulting ethical rules or moral principles. Instead, the organisms would be able to make such decisions by using their modelling capacity to identify the actions that would produce the best evolutionary outcomes. Broadly, this would equate to the behaviours that would have prevailed if normal evolutionary selective processes had continued to operate. For such an organism, ethics and morality would be abolished as primary categories to guide behaviour.

This outcome can be seen as the culmination of a trend that began when organisms first started to develop a capacity for mental modelling. As we have seen, mental modelling enables an organism to identify the future consequences of its actions. It can therefore be used to discover effective adaptations without the costly trial-and-error involved in natural selection or trial-and-error learning.

However, when a capacity for mental modelling first emerges, its ability to anticipate future events is limited.

As the capacity for mental modelling improves, the organism will be able to take into account the impacts of behaviours that are more distant in time and space. Increasingly, the organism's behaviour will be shaped by mental models, rather than by inherited genetic or cultural predispositions. Wherever this replacement occurs, the organism will experience its decision-making as being of a practical nature, directed at identifying actions that produce particular adaptive outcomes in the world. Increasingly, decision-making will use mental models to identify actions that will achieve the organism's goals. No longer will these decisions be motivated primarily by ethical and moral considerations.

As this capacity continues to develop, the organism will eventually be able to construct mental models of the longer-term evolutionary consequences of its actions. Even actions that have highly complex and longer-term impacts will now be able to be evaluated effectively by the organism using its mental models.

The organism will also experience these evolutionary decisions as practical matters, not as ethical and moral choices. The choices that it makes will be determined by predictions about the evolutionary consequences of the alternatives, not whether they are right or wrong according to some culturally-inherited religious or moral system.

For an example in human evolution, consider the choices that are made about what food to eat. Most humans have genetic predispositions to enjoy sweet foods. This predisposition was likely installed in humans when sugar and other carbohydrates were in short supply in their environment. However, in modern environments this is not the case, and this predisposition often drives obesity and other health problems. Our mental models enable us to understand this, and we can, as a matter of practicality, consciously decide to restrict our sugar intake.

In the past, some food choices were also guided by religious beliefs. For example, a number of religious systems prohibited the consumption of pork, often for reasons that had nothing to do with the known effects of eating it. Increasingly, however, food choices are now decided as matters of practicality, based on science-based mental models.

These examples also illustrate another feature of this major transition in evolvability. It is obvious that the transition can be made only once an organism has developed the cognitive capacity to construct and operate mental models of complex future circumstances. However, it also requires organisms to be able to free themselves from the dictates of their genetic and cultural predispositions and conditioning.

It is not enough just to know what to do to thrive and survive. It is insufficient just to be able to see that particular predispositions are maladaptive in current circumstances. Knowledge alone does not free individuals from inculcated predispositions. They need to develop an ability to move at right angles to pre-existing predispositions. Many humans know that excess sugar and fat in their diets is harmful to their future health and survival. However, many are unable to override the predispositions that drive their over-consumption.

Recapping: Improvements in evolvability will eventually produce organisms that have developed a comprehensive understanding of the evolutionary processes that have shaped them and that will determine their survivability into the future. The organisms will be able to use this understanding to identify the trajectory of evolution and to locate themselves and their social systems along it. This enables them to see how they need to adapt and evolve so that they remain aligned with the trajectory in the future, and are not selected out of existence.

Increasingly, they will be able to use this evolutionary understanding to make decisions about how to evolve themselves and their societies. No longer will they blindly follow ethical, moral, and religious principles that governed the behaviours of their ancestors. To do so would be absurd, particularly where the strategies that they derive from their mental models conflict with these predispositions.

Of course, on this planet at this time, humanity is at the threshold of making this major evolutionary transition in evolvability. At present, almost no humans have reached the stage where their cognitive capacities are sufficient to build mental models of enough scope and complexity to envisage how we will need to evolve and adapt to survive indefinitely into the future. However, we have developed the cognitive capacity to construct analytical/rational models.

As we have discussed, analytical/rational thinking underpins current mainstream science and has driven technological development. But it is limited. It builds only simple, mechanistic models that can be 'thought through' analytically. Consequently, it is only useful for modelling those limited parts of reality that can be approximated by simple, mechanistic models. It is next to useless for building models of most of reality. Much of reality is too dynamically complex to be represented adequately by such models. As a result, analytical/rational thinking is largely incapable of building and operating models of complex evolutionary processes.

At present, for nearly all humans, ethics and morality have not yet been ousted by evolutionary modelling. However, religious systems (and the ethics and morality that they entrench) have been seriously undermined by the rise of analytical/rational cognition. Religious systems have been unable to defend themselves effectively against critiques of religious beliefs that have been made by analytical/rational thinkers. As Nietzsche pronounced, God is dead, and we, the rational, have killed him.

As Nietzsche also demonstrated, this destruction of the systems of religious belief that had previously provided a strong basis for ethical and moral systems, puts humanity in a very difficult and dangerous position. When religion ruled, it reinforced moral and ethical principles that assisted humanity to adapt effectively, including by producing cooperative social systems. Unfortunately, in order to understand the wisdom of many of these principles, cognition that is capable of modelling complex dynamical systems is required.

As a result, when rationality killed God and undermined the moral principles that religions had entrenched, analytical/rational thinking was not up to the task of developing replacement principles from scratch. This was particularly the case for the pro-social principles necessary to bind together humans into complex social systems. Analytical/rational models are incapable of understanding what is required to accomplish this.

Tragically, humanity was left without the wisdom needed to replace what it had destroyed. As Nietzsche pointed out, what humanity needs, therefore, is what he referred to as a re-valuation of all values. We need to reassess all our values and goals. Consciously and intentionally, we need to construct the mental models needed to systematically re-evaluate our goals and strategies, particularly in relation to our social systems. We need to change them as necessary to adapt ourselves and our social systems to our current circumstances, and to evolve them as circumstances change.

If our new goals and strategies are to be viable, they will need to be able to withstand rational criticism. They will need to be science-based. But to undertake this re-valuation effectively, we will need to develop what I am referring to as metasystemic cognition—the ability to construct and operate mental models of complex, evolving, interacting, dynamical systems. However, at the present time, we do not have these higher cognitive capacities. We have killed God, but do not yet have a replacement. We are facing existential threats that, at least in part, have been generated by this failing. We are currently in a race between a descent into self-destruction and the getting of wisdom in the form of metasystemic cognition. The fundamental purpose of this book is to assist in the development and spread of metasystemic cognition across humanity.

In order to implement the results of this re-valuation so as to complete this major evolutionary transition, humanity will also need to develop the capacity to be self-evolving, both as individuals and collectively. We will need to transform our psychology so that we are able to change ourselves and our behaviour at will. This will require the development of practices that free ourselves from our genetic, cultural, and social predispositions and conditioning.

Broadly, the same applies to organisations of humans, including our societies. They too will need to adapt using appropriate metasystemic cognition, and to become self-evolving entities in their own right.

I have already outlined my earlier fumbling and stumbling attempts to identify practices that are capable of scaffolding a capacity for self-evolution. There is more about these efforts later in the book. As my thinking about the nature of major evolutionary transitions developed, it propelled me further in the direction of discovering practical methods for becoming self-evolving.

It is important to emphasise a number of points about this major transition in evolvability:

First, it does not commit what is known as the naturalistic fallacy. This fallacy argues that it is invalid to derive an 'ought' from an 'is'. In other words, it is illogical to argue that humans ought to do something solely on the basis of facts about the way the world is. In particular, the naturalistic fallacy has often been used to attack attempts to use evolutionary theories to suggest what we should do with our lives. For example, it can be legitimately argued that just because evolution might have favoured aggressive competition (or cooperation), it does not

173

follow that humans ought to do likewise in their lives. The fact that evolution appears to favour something does not mean that humans ought to.

But what I am proposing here does not suffer from this deficiency. It derives its 'oughts' from other 'oughts' in combination with relevant facts, not solely from facts. There is no logical fallacy involved in deriving 'oughts' from other 'oughts'. For example, if an individual holds a particular value, it is perfectly rational for the individual to use that value to derive new values that are consistent with it. This is clearly the case where satisfaction of the derived values will lead to the satisfaction of the original value.

The use of relevant factual information in this derivation of new values is also perfectly legitimate. Particular facts might be highly relevant to identifying the circumstances in which the pursuit of the new value is consistent with the pursuit of the original value.

Individuals who become self-evolving do not commit the naturalistic fallacy—they use evolutionary theory and model building to choose their actions only to the extent that this is consistent with their most fundamental, pre-existing values. In effect, they rely on arguments of the sort: "If you want to ensure that humanity survives under these particular circumstances, you ought to adopt the following kind of goals, and take the following kinds of action ..." Of course, if a person has no preference about whether humanity survives or not, these 'oughts' would not apply to them.

A narrow analytical philosopher might attempt to counter such an argument by suggesting that it necessarily leads to an infinite regress—the argument cannot ground any 'ought' at all without relying on some pre-existing 'ought'. However, this ignores the fact that any real-world human being who engages in making such a decision already comes fully equipped with plenty of pre-existing 'oughts'. There is no such thing as a functioning 'oughtless' human being. Show me an 'oughtless' human, and I will show you a dead body. This fact terminates any regress well short of infinity.

Furthermore, nearly all humans want civilization to continue to function effectively for at least as long as their lifetime. This goal is capable of generating a plethora of 'oughts' about how we need to adapt and evolve within that timeframe. This is because if human civilization is to survive and thrive this century, individual humans and their societies need to evolve in many particular ways in the coming years.

This brings me to the second point that I need to make about the transition to conscious self-evolution. The evolution that I am talking about is not a process that occurs only in the far-flung future. It is true that complex adaptations took millions of years to emerge when the evolutionary process depended on genetic mutations that were subjected to selection over many generations. Gene-based natural selection tends to search possibility space very slowly. Many billions of organisms must be selected out of existence for these mechanisms to discover and build complex adaptations.

But to a large extent, these gene-based mechanisms have been superseded as evolvability itself evolved and improved. Gene-based mechanisms do not generate most of the evolution that occurs at the human level now. As the great philosopher of science Karl Popper put it, our ideas now die in our stead. We now try out possible adaptations in our heads. Rather than taking millions of years, evolution can now proceed as fast as we think. Significant changes can take days or even hours to emerge and begin to spread.

I used to joke with my two daughters that they were not my real children. I would tell them that from an evolutionary perspective, my real children are my ideas.

Evolution is occurring now, as you read this page. As we will see in more detail, there are numerous actions that individuals need to take here and now if humanity is to survive and evolve successfully into the future.

Up to this point, evolution on Earth has proceeded largely blindly and without foresight. After this transition, it will proceed consciously, anticipating how evolutionary demands are likely to change in the future. As Marshall McLuhan pointed out, past evolution has been like driving a car while looking only in the rear-view mirror. Previously, evolution proceeded by taking into account only the past consequences of actions.

The acquisition by humans of metasystemic cognition, together with a capacity to self-evolve, will fundamentally change the evolutionary process on this planet. And it will transform not just the role of ethics and morality in shaping human behaviour.

In large part, our desires and motivations, not just those associated with our ethics and morality, have been shaped by our evolutionary past. Although the means for satisfying our desires have changed enormously, we continue to pursue much the same proxies for evolutionary success as our ancestors did fifty thousand years ago. We spend our lives chasing the positive feelings produced by experiences such as popularity, self-esteem, sex, feelings of uniqueness, power, eating, and social status. We strive to avoid the negative feelings that go with experiences such as stress, guilt, depression, loneliness, hunger, and shame.

In effect, we live in a virtual world created by past evolution. What we take to be important and valuable is an illusion created by evolution to control our behaviour.

This understanding resonates with the suggestion made by some Eastern spiritual traditions that the experiences that we take to be real and important in our lives are, in fact, Maya. They are illusions that orchestrate our behaviours until we wake up to reality's true nature.

It is also consistent with Plato's allegory of the cave, in which humans mistake shadows on a cave wall for reality but are chained by their desires to see only the shadows.

As Shakespeare put it, an individual: "struts and frets his hour upon the stage, and then is heard no more. It is a tale told by an idiot, full of sound and fury, signifying nothing." The 'tale told by an idiot' is the story we tell ourselves and others about our life in Maya.

If humans complete this transition, it will represent a major evolutionary transition in evolvability. It will be of evolutionary significance, not just of historical importance. This transition will enable us to free ourselves from the dictates of our evolutionary past. A new kind of human will enter history and evolution. It will change everything.

15.

Working on Awakening

The writing of my book 'Evolution's Arrow' went well during 1999. As I do, I began by investigating techniques for writing books, and adapted them to my circumstances and needs. I found that writing for about four hours a day for four days a week was enough to produce a good first draft by the time I had to return to my paid work. I would then be able to edit and polish the book when I resumed work.

Consequently, I had plenty of spare time at home while I was writing the book. My daughters were teenagers, one 16, the other 14. Usually, I had finished my writing for the day before they arrived home from school. Every afternoon, we would all get together as a family for a talk and a laugh when they got home. In general, I was extremely happy being married and I found having two bright and lively daughters truly magical. But this went to a new level during my year at home. It was the happiest year of my life. Producing a book was a bonus.

Finishing writing around 3 pm also gave me the opportunity to get involved in the girls' sporting activities. Australia is very sports-focused, given its warm weather and outdoor lifestyle. And both my wife's and my own family had a history of involvement in sports. My eldest daughter, Anna, was a talented cricketer. She specialised in batting, and her stroke-making was elegant. Eventually, I became the manager of her cricket team. Other fathers would often compliment me on her batting style, thinking that I had coached it into her. But I had nothing to do with it. It was largely innate.

Her batting skills and elegance were very surprising. When she was younger, she was nicknamed 'Unco-Anna' at school. She was visibly uncoordinated when running and catching balls. When she asked me if I could take her to play cricket, I would make excuses. I put it off for years. I thought that I was protecting her from the embarrassment of dropping a key catch in a game, causing her team to lose. But when she picked up a cricket bat, she was transformed. I still feel a bit guilty for not taking her to cricket at a much younger age. Fortunately, however, this was the only mistake I made during Anna's upbringing.

My other daughter, Libby, had natural coordination skills and athleticism in abundance. She took to playing tennis. There was something very enjoyable about going with her to the local tennis courts in the afternoon and taking her to tournaments on the weekend.

Because we lived in the Australian Capital Territory with its relatively small population compared with other States, it was relatively easy to get selected to take part in annual State-level competitions. You did not have to be very serious about your sport and have it as the main focus of your life. Anna represented the ACT in cricket for several years, and Libby represented the ACT in tennis.

My involvement with Libby's tennis also played a significant role in my further exploration of practices that could help to scaffold capacities for self-evolution and self-mastery.

When Libby began playing some tennis games at night, she found that she could not serve with her normal effectiveness. On some occasions, she even missed hitting the ball after throwing it in the air to serve. This was inexplicable to her coach and me. She had no trouble at all hitting volleys that were travelling towards her at a far greater speed than the ball did when she was serving. But gradually, we were able to piece together the cause from a number of seemingly unrelated events.

One of these occurred while Libby was practicing at a local tennis court, and rain forced the players indoors. Libby and her friends began playing table tennis. But Libby found that she often missed hitting the ball back. Again, this did not seem consistent with her ability to hit tennis balls from all angles and speeds.

Then her mother remembered taking Libby for an eye check when Libby was young. The doctor told her that Libby would never pilot a fighter jet. He explained that she had a 'lazy eye'. It became apparent that Libby was playing tennis with only one good eye. This was confirmed when she tried covering up her normal eye. Her remaining vision was very blurred. As well as never becoming a jet pilot, Libby would never be able to become a top-level tennis player.

This intrigued me. I had learned at school that binocular vision was required for depth perception. Yet Libby played tennis at a very high level for her age. How could she possibly do this without effective depth perception?

I used several techniques to test whether or not Libby had functioning depth perception. For example, I tried her out on stereograms. These are two-dimensional pictures that, when viewed in a particular way, produce perceptions in 3D. However, this did not provide a definitive answer. The stereograms that I could find did not work very well for me, and I had two normal eyes and good depth perception.

I looked for another method. At the time, it was winter in Canberra, and the elm trees were without leaves. After a while, I discovered that if I looked long enough into the bare branches and twigs of a tree, I would be able to see the tree in three dimensions. Branches that were nearer to me would 'pop out' from branches that were further away. The effect was striking.

It took some practice, but eventually I found that I could experience my entire visual field in vivid 3D. I was able to see trees as a whole, avenues of trees as a whole, people and crowds of people in 3D, and buildings in 3D. When seen as a whole, trees took on an entirely different character. Somehow, their trunks looked like crystals, fixed in space, embedded in stillness.

Finally, I understood a line from a Joyce Kilmer poem that I had memorized for English exams many years ago: "I think I shall never see a poem as lovely as a tree". I could now see what he saw when he looked at a tree in a state of stillness and presence.

When I looked at the world the way I had learned to see the leafless branches of Canberra elm trees, everything looked beautiful and extraordinary. As the great Italian film director Federico Fellini said: "If you see with innocent eyes, everything is divine." Libby's problem had led me to discover how to look at the world with innocent eyes.

Libby's lack of depth perception could not be remedied at her age. If she had received appropriate treatment when young, her brain could probably have adapted to overcome the issue. But as a 14-year-old, it was too late.

Nevertheless, she was still a very good tennis player. Apparently, with practice, her visual system had learned to use other cues that enabled her to hit the ball successfully without binocular vision. Presumably, for example, her perception could use factors such as the size of the image of the ball falling on her retina. Conceivably, this enabled her to estimate where the ball was located in space relative to her and to hit the ball reliably.

It was not immediately obvious to me what was happening when my attempts to see my external environment in 3D somehow produced a state in which the visual world appeared more spacious and vivid. But my experience seemed to tally with how Gurdjieff responded when asked what it was like when you awaken in the midst of ordinary life. "Everything more vivid," he said.

Before long, I could enter this state at any time I wanted to, whether I was walking in a park or in a city, or during a meeting at work. It seemed to be a major step forward in the development of a capacity to be awake in the midst of ordinary life. By this time, I had been meditating regularly for 8 or so years, often with eyes open in public places. Whilst meditating, I could enter a state of internal stillness and spacious awareness. But this had not translated into a capacity to be awake while I was engaged in daily activities.

Now I could enter this state at will. But this only got me into the state if I actually remembered to decide to enter it. Often for long periods, I would forget all about entering the state. When I was caught up in interactions with people or in demanding work, my awareness tended to be embedded in these activities, and would remain embedded. There was no surplus attention left over that could remind me that I wanted to return to being awake as soon as possible. As is the case for everyone, when I was thinking deeply about something, the rest of the world disappeared from my awareness. I was totally absorbed in the thinking.

It was obvious to me that I still had a long way to go to awaken fully and to use this capacity to become self-evolving in ordinary life. Much work on myself was needed to take these further, important steps. But I could see the potential. When I was awake during normal activities, I tended to be unabsorbed in thinking and feeling. While I was in this state of spacious awareness and stillness, thoughts, and feelings would still arise to some extent, but I was less likely to become absorbed in them. I was less likely to find that my awareness had contracted down and was now absorbed in particular thoughts.

In the state of stillness, thoughts and feelings were less likely to arise. But this was not because they were actively suppressed. In fact, when they did arise, I would experience them more fully and vividly, just as I experienced the external environment more fully. In this state, trees and other objects did not automatically capture and contract attention. I could move attention to them or away from them at will. It was now much the same for thoughts and feelings.

Towards the end of 1999 and during the early 2000s, my major goal became to explore waking up, self-evolution, and associated issues in depth. I was propelled along this path by a combination of two developments: what I had learned from my exploration of Libby's tennis issue; and what my evolutionary thinking was increasingly revealing about the critical importance of a capacity to awaken in ordinary life.

In the meantime, I had finished the final draft of Evolution's Arrow. I was not optimistic about finding a publisher. I did not have the public or academic profile needed to reassure publishers that the book would be a success. I could spend years trying to find someone to publish the book without much chance of success.

For these reasons, I imposed on myself a deadline of six months to find a publisher. During that time, I had a couple of near misses, but nothing concrete. When the deadline passed, I proceeded to self-publish by putting HTML and PDF versions of the book on the internet so that it was freely available.

I followed this up by spending considerable time bringing the book to the attention of relevant academics and thinkers. For example, I combed the records of international conferences on systems science to find people who had presented relevant papers, googled their email addresses, and sent them materials and links about the book.

At the time, I thought that it was possible that some of the evolutionary thinkers that I contacted would tell me that my evolutionary theories were not novel. They might suggest that my ideas had already been developed and published by others. This was a possibility that my previous research could not rule out. Prior to writing Evolution's Arrow, I had almost no direct contact with other evolutionary thinkers. I had not attended any international conferences, and my papers had received little attention in the literature, positive or negative. Consequently, I could not be sure whether my ideas had been arrived at previously by others.

Furthermore, as I have mentioned, by now the theories developed in the book seemed obvious to me. I thought that its central ideas would also have been obvious to others and been developed previously.

I was surprised when no one contacted me to say that my theories had already been published by others. Nor did I receive any negative responses about the ideas. In general, the responses were positive and encouraging.

Perhaps the most significant response was from the internationally renowned cybernetician and systems thinker, Francis Heylighen, a lecturer at the Free University of Brussels. He was well known for his work on the Principia Cybernetica Project. This was an ambitious attempt to post comprehensive materials about cybernetics on the internet and make them freely available and accessible. He liked Evolution's Arrow, and made a PDF of the book available through the Principia Cybernetica Website.

Francis also invited me to become affiliated with his Department at the Free University of Brussels. He said that he had noticed that I did not have any formal academic affiliation (in fact, in some earlier papers, I had given the Department of Industrial Relations as my address, somewhat in jest). Francis suggested that a proper academic affiliation would make it easier to get future papers published. I jumped at his kind offer. A few years later, Francis established a formal research group that

I joined: The Evolution, Complexity and Cognition Research Group, at the Free University of Brussels. A perfect fit.

My goal of exploring the capacities associated with awakening and self-evolution had two interrelated purposes. Neither had anything to do with spirituality. As I have indicated earlier, I had no interest or belief in religious, supernatural, or mystical phenomena. This had not changed. However, none of this prevented me from learning from the discoveries of the contemplative and spiritual traditions, and attempting to understand the states that they explored, but from a materialist perspective.

The first purpose was to discover practices and knowledge that would enable me and anyone else to awaken permanently in ordinary life. However, for me, awakening was a means to an end, not an end in itself. I was interested in awakening primarily due to its potential to enable self-evolution. Furthermore, awakening had the potential to enable an individual to see one's thinking as object. As I have mentioned, this was critically important for enabling an individual to see the limitations of their existing thinking, and to enhance it intentionally and recursively.

In addition, I also had some experiences that indicated that awakening could provide access to cognitive resources that were important for creative thinking. These include intuitions and pattern-recognition capabilities that appear to come from the unconscious mind, but are normally excluded from awareness by incessant absorption in thought.

In short, awakening seems to have the potential to enhance cognition and to enable self-evolution.

My second purpose was to develop a scientific theory of the development of the capacities that enable awakening and self-evolution. As I have mentioned earlier, I wanted to develop a materialist model that could explain, for example, why it is that humans can achieve changed states of awareness; how and why these can be produced to some extent by meditation; how awakening can be produced; what enlightenment is and how it can be achieved; how spiritual practices produce the effects that they do; and how can one free oneself from the dictates of genetic, cultural and social predispositions (including conditioning); and so on.

As I have discussed previously, such a theory and the models that accompany it would enable new and improved practices to be invented and refined. It would transform the field of spiritual and contemplative practices into a set of technologies. In the process, practices would be shorn of spiritual and religious trappings. Practices would be optimized so that they are able to produce capacities such as awakening, enlightenment, and self-evolution in the most efficient and effective ways.

These practices and states would no longer be associated only with religious and spiritual traditions. Instead, they would be seen as potentials in all human beings that exist because of the particular way in which the human brain and psychology are organised.

As happened soon after other domains of human experience were first subsumed into science, the techniques embodied in these practices would likely undergo rapid technological development. And just as other technologies are continually improved and optimized, so too would these psycho-technologies.

At about this time, I came across an advertisement for a group in Canberra that called itself 'The School for Self Knowledge'. The advertisement invited people to participate in a series of lessons about the meaning of human existence and associated issues. No background was provided about the history of the group or the sources of the ideas that it presented. However, given my goals and interests, I had to go and find out if they knew anything that could help me to achieve my goals.

I went to the School for Self Knowledge regularly for two years. Early on, I recognised that some of the ideas it presented were unmistakably Gurdjieffian. But other ideas were not. They taught a simple practice that they called 'Being here now'. It involved relaxing and giving surrendered attention to each of the senses in turn. During the practice, if you found that you were thinking or emoting at any time, you returned your attention back to the sensations. The practice had the effect of stilling the mind and bringing the practitioner into the present.

I already had an effective meditation practice and had developed the ability to come into the present at will in the midst of ordinary life. But the school held out the promise of other, deeper practices if you continued with them.

I found the lessons useful. Irrespective of the justification they gave for the ideas they presented, the ideas were interesting to contemplate. I would sit in the lessons in a state of presence, and reflect on what they had to say.

Gradually, I began to see that common principles underlay the practices that I used, the practices that they advocated, and others that I had read about. I began to make progress in developing a model that could explain objectively, non-spiritually, how these kinds of practices produced particular psychological effects, including awakening and presence.

Eventually I realized why it was that when I looked up into leafless elm trees and attempted to see them in 3D, I would come into the present in spacious awareness. In order to see the trees in 3D, I had to relax and still my mind. I had to give bare, surrendered attention to the branches. When I looked with a still mind, things popped out in 3D. Colours, shapes, and textures were more vivid. The whole processing power of my mind was focused only on sensing. None was taken up with thinking. Consequently, my awareness was more spacious, and my perceptions were more detailed and vivid. I knew the meaning of the statement: "Be still and know that 'I am'."

I noticed that when I came into the present at will, my mind was still. I was looking without thought or judgment. This is what I mean when I refer to 'bare attention' and 'surrendered attention'. I also noticed that when I started thinking, my spacious awareness contracted down, and I became embedded or absorbed in thought. I was no longer in spacious presence.

It became obvious to me that similar principles were responsible for the effects of the school's 'being here now' practice, and many others I had read about. Giving bare, surrendered attention to sensations, and bringing attention back to them whenever you realize that you are re-embedded in thought or feeling, will still the mind. This stillness of mind will occur whether the focus is on sensations of the breath, feelings in one's legs, sensations within your head, the feelings of air against your skin, or any other sensations, including sensations that are imagined or visualized.

The key is this: humans can come into the present if they rest attention on something that occupies and attracts their attention, but that does not itself evoke thinking or feeling. By giving attention to something that does not itself evoke thinking or feeling, an individual is able to rest attention on it for a longer period without interruption, and the more still the mind will become.

A related form of practice is to begin by surrendering all goals, intentions, and motivations. Then, when one finds oneself re-embedded in thought or feeling, one returns to the surrendered, intentionless state. And so on. When one is surrendered, thoughts and feelings will tend not to arise. The mind will become still. This approach can often prevent re-embedding more effectively than can the method of resting bare attention on sensations. This is because it goes beyond occupying attention with an alternative to thoughts and feelings. Instead, the intentions and motivations that would otherwise generate thoughts and feelings are dropped.

These two methods can be combined in various ways.

It is also worth emphasising that both these methods utilize the fundamental principle that I have described previously as being one of the pillars of meditation practice: when one finds that one has become re-embedded in thought or feeling, one moves one's attention gently and non-judgementally back to attention and/or surrender. Done repeatedly, this will still the mind and strengthen the capacity to dis-embed and to remain dis-embedded.

Increasingly, I came to understand that these principles underlay many of the practices that had been found to be capable of training the ability to enter spacious presence on an ongoing basis. When I examined other methods, such as the use of mantras, it was generally easy to see how they produced their effects. The use of mantras, for example, can exclude thinking by fully occupying attention while also providing an 'inert' stimulus.

Progressively, I integrated these experiences and insights into a scientific model of the practices and the states they could produce. Initially, when I learned how to come into spacious awareness at will, I was unaware of how I achieved this. I did not know consciously what my mind or body was doing when I underwent this change of state. But my model building had now provided me with an understanding of what was going on.

This conscious understanding enabled me to work out how to modify what I was doing in order to enhance its effectiveness. It also led me to develop a general understanding of spiritual practices, to see their common features, to distinguish between elements of practices that contributed to their beneficial effect and elements that were just spiritual window-dressing and that could be dispensed with, and to identify how practices could be improved and optimized.

I put these insights to work in my evolutionary theorizing. The model I had constructed had important implications for the design of practices that would enable individuals to become self-evolving. In particular, from my own experience and from the predictions of the models, it was clear that in a state of spacious awareness, an individual was dis-embedded from and non-attached to thoughts and feelings. In such an awakened state, the individual experienced their thoughts and feelings as objects that arose in their awareness and then dissipated. Thoughts and feelings were now objects in the same sense that external phenomena often tend to be objects in our visual awareness—we can give them attention if we choose to, or take our attention away from them. Whether we become embedded in them is a matter of choice.

Once we become non-attached to thought and emotions, they no longer control our behaviour. They do not automatically contract our awareness. If we choose to, we can watch them arise and then dissipate. In an awakened state, we can remain in spacious awareness, no longer jerked around by thinking and feeling. This enables us to choose consciously to move at right angles to our emotional predispositions, if we decide it is appropriate to do so.

This is the central capacity needed to become a self-evolving being. One must be able to free oneself from the dictates of one's genetic, cultural, and social past, including from one's conditioning. This frees the self-evolving being to do whatever is required to satisfy the demands of future evolution. Instead of being continually waylaid by emotions and other predispositions established by past evolution, the self-evolving being can behave in whatever ways are necessary for survival and thrival into the future. As circumstances change and change again, such individuals can continually adapt and evolve their behaviour.

Developing these capacities is essential if an individual is to pursue evolutionary goals successfully. As I have mentioned, it is not enough to develop the ability to construct and use complex models to identify what one must do to achieve evolutionary goals. Higher cognition is not sufficient. One must also have the psychological capacity to actually implement the required actions, despite the fact that they may conflict with the individual's pre-existing genetic, cultural, and social predispositions and conditioning.

In addition to freeing themselves from conflicting predispositions, self-evolving individuals must also intentionally develop ways to find motivation and satisfaction in whatever they need to do to achieve their evolutionary goals. The capacity to enter presence at will is central to achieving this. Presence and awakening are the royal road to finding enjoyment in whatever one chooses to do.

Individuals can use an ability to manage their emotions and motivations in order to support their pursuit of evolutionary goals. At present, most individuals are able to use their control over their body to adopt a physical posture that is conducive to undertaking a particular task. Self-evolving individuals are also able to adopt an emotional posture that provides the energy and motivation needed for particular pro-evolutionary tasks.

Presence also facilitates the use of additional techniques such as re-conditioning oneself, re-framing one's perspectives, and using visualization methods, including active imagination. By employing these kinds of approaches, individuals can develop the ability to align all their psychological goals with evolutionary goals and continue to do this as circumstances change. Those who do not will continue to spend their lives pursuing proxies for past evolutionary success that may have been rendered maladaptive by changed circumstances.

With these capacities, self-evolving beings who pursue evolutionary goals never have to be psychological altruists or psychologically self-sacrificing.

I developed these ideas in greater detail in an academic paper published in 2001, titled 'Future Psychological Evolution.'[11]

After two years, I left the School for Self Knowledge. I had undertaken some research into its history, and discovered where most of its ideas and approaches came from. They had an interesting backstory. Initially, their main influence was the work of the Russian thinker Pyotr Ouspensky who had been a follower of Gurdjieff. After leaving Russia following the revolution, he taught Gurdjieffian ideas in England for many years and attracted a large following. He was an exceptional writer, and produced the most well-known, popularized account of Gurdjieff's early teachings.[12]

But Ouspensky himself never seemed to really 'get' the core of the teaching. He was very much intellectually orientated, and seemed to think that awakening could be understood and achieved intellectually. He never appeared to fully realize that Gurdjieff's teachings were primarily about attaining new psychological capacities and different states of consciousness, not about new ideas. It was not possible to think one's way into these higher states. In fact, they necessitated the stilling of one's mind and becoming less attached and identified with ideas and thinking. Ultimately, for the new kind of human being that Gurdjieff pointed to, it was the stilled mind that would take control of thinking, deciding when thinking would be utilised, and for what purpose.

However, Ouspensky was not the only person or group to influence the School for Self Knowledge. When Ouspensky died in 1947, he left his followers without a leader. Perhaps more significantly, he also left them without a comprehensive set of explicit awakening practices. This led some of his more resourceful followers to travel the

---

[11] Stewart (2001) – see References
[12] Ouspensky (1949) – see References

world to find suitable practices that had been developed by other contemplative and religious systems. Apparently, one persistent group encountered practitioners of Advaita Vedanta on the banks of the Ganges River in India, and thought that they had found what they were looking for. This group passed their discoveries on to the couple who had founded the School for Self Knowledge in Canberra.

Eventually, I discovered that the founders of the school had formed close links with a particular high-ranking Vedanta teacher/leader in India. Each year, the leaders of the school went to India to visit him. Soon after I discovered this, the school announced that students who had been with the school for as long as I, were invited to undertake some kind of initiation ceremony. Apparently, this involved making various commitments, including agreeing to tithe one's income for the benefit of the school. The leaders of the school began to refer to the Indian Advaita Vedanta teacher as 'His Holiness'. That was enough for me. They seem to have misunderstood completely Gurdjieff's ideas and Advaita Vedanta. I wished them the best of luck, and left.

During the next few years, I continued with my meditation and awakening practices, and experimented with trying to improve them in various ways. In order to remind myself to wake up often throughout my working day, I used my email system to automatically send me messages. I also put statements and relevant quotes from the traditions on my screen saver, and used events that recurred during a normal day at work as reminders to wake up. For example, I used work meetings as a reminder to come into the present and remain present. It was fun to find that I was often the only person who was awake during a meeting, in uncontracted awareness. Of course, if you are asleep and embedded in thoughts and feelings, you will be oblivious to whether others are awake or asleep.

I found that being present when interacting with others made me considerably more aware of their reactions and non-verbal behaviours. I noticed far more about their emotional states and motivations. I could put these additional insights to work in achieving whatever goals I had. I got even better at manipulating meetings. I also found that when I was present, my thinking was often clearer, sharper, and more insightful.

This was particularly the case if I had been under pressure and was somewhat stressed.

However, I did not have any significant emotional or other psychological challenges at the time. Both my work and family life were easy, and I was generally happy. I was very good at using my intelligence to control my environment and avoid difficulties. My evolutionary theorizing was going well. I was manipulating my work demands successfully so that I had plenty of time to do what really mattered, and my wife and daughters seemed happy and well.

I did not have any particular fears or traumatic experiences. However, my work did require me to do things that many would find stressful and challenging. For example, as mentioned, a central requirement of my job was to develop and then present the Government's case to argumentative arbitration tribunals. However, I tended to enjoy this.

Furthermore, I was obviously left-brain dominant, and somewhat stunted emotionally. Consequently, I tended not to be beset by strong negative emotions. Of course, this had disadvantages as well as advantages. As we shall see, emotional functioning plays a significant role in the development of metasystemic cognition. In part, my experimentation with meditation had originally been motivated by the goal of becoming more aware of my emotions, and using them for cognitive purposes (for example, emotions utilize pattern recognition capacities, which are important for the functioning of higher cognition).

In retrospect, however, it is clear that my ability to manoeuvre around difficulties also had downsides. I have mentioned that my refusal to accept promotion to the level of Senior Executive had enabled me to spend much more time doing my evolutionary work. Furthermore, it enabled me to avoid the stress that I would have encountered as a perfectionist doing a job that could not be done to my standards in a reasonable time frame.

However, if I had taken the promotion, I would have been forced to confront my perfectionist demons. I would have had no option but to use presencing techniques in order to fully accept and surrender to the negative feelings that would have initially been generated by being forced to compromise my standards. Without the time to re-write the work of everyone in my Branch, I would have had to deal with being responsible for work that I knew to be incompetent. I had successfully avoided what would otherwise have been a powerful driver of my further development.

16.

More Publications and More Awakening

In 2004, I came across printing-on-demand as an effective and cheap method of making Evolution's Arrow available in book form through Amazon and other online booksellers. Using this system would enable hard copies of the book to be distributed internationally. I got hold of relevant software, typeset the book, designed a cover, and uploaded it to the print-on-demand provider. Under this system, when someone ordered a copy of Evolution's Arrow from Amazon, Amazon placed an order for one copy with the provider, and they printed a single copy of the book using an electronic file.

My wife and I had begun to go on some overseas trips now that our daughters were old enough to look after themselves. We were in Hanoi when I asked her to wait while I went into an internet café to check whether Evolution's Arrow had yet been made available on Amazon. I found that not only was it there, but it was the highest selling book at that time on Amazon. I was sure that this was a mistake.

Then when I checked my emails, I found one from Michael Dowd, an American popularizer of an evolutionary worldview. His email praised Evolution's Arrow enthusiastically and indicated that he had just recommended it strongly to the thousands of subscribers to his newsletter. This explained the large number of sales on Amazon that day, and why they fell off a cliff very soon after.

I was to get to know Michael over the coming months and years. He described himself as America's evolutionary evangelist. Together with his partner Connie Barlow, he travelled the United States in a campervan, giving talks about an evolutionary worldview, mostly to church groups.

Broadly, Michael set out to re-interpret religious teachings from an evolutionary perspective. In his talks, he demonstrated that far from undermining religious beliefs, an evolutionary worldview could enrich and deepen them. For example, he argued that to live 'in right relationship' with God equated to living in right relationship with the whole of reality. This in turn entailed living in an environmentally sustainable way, and living in alignment with the trajectory of evolution.

Soon after his initial contact, Michael told me that the evolutionary vision outlined in Evolution's Arrow had inspired him to organise a meeting to discuss and promote the evolutionary worldview it gave rise to. He said that he would be able to make sure that the meeting was attended by leading academic evolutionary scientists. He would also invite spiritual progressives and other progressive thought leaders who had begun to consider an evolutionary worldview. He hoped that I would be able to join the meeting, but if I could not, he would proceed with it anyway in 2005.

Taking a couple of weeks off work (officially) was not a problem, and I thought that it would be an interesting experience, whoever turned up. I had to go.

The meeting, now referred to as the Evolutionary Salon, took place in May 2005 in California. The leading evolutionary scientists that Michael invited had agreed to attend initially, but all dropped out as the Salon approached. This was not because Michael lacked influence amongst evolutionary scientists. As I increasingly discovered, Michael had knocked on the door of the most prominent evolutionary scientists in America, and had generally impressed them with his knowledge, enthusiasm, and authenticity. However, it was becoming clear that the Salon would not be anything like a traditional scientific conference. This appeared to worry the academics. One of the greatest fears amongst academic scientists is that their work will be seen to be associated with fringe ideas and crack-pot science. It was much safer for them to avoid these kinds of risks by cancelling their attendance.

The Salon was a fascinating experience.

My main goal for the Salon had been to meet academic scientists and discuss with them the evolutionary theories that I had published. I wanted to know what they thought about my ideas and the reasons they had for accepting or rejecting them. It was not that I wanted reassurance about my theories. By then, I was confident about the novelty and the importance of my ideas. However, if any individual scientist rejected my ideas, I wanted to know in detail the reasons. This would enable me to design strategies to overcome any resistance to my theories among evolutionary scientists.

I knew that an evolutionary worldview was dead in the water if the science that underpinned it had not received mainstream support. This was the case irrespective of the validity of the science from which it was derived. Any attempt to publicize an evolutionary perspective that had significant and controversial implications for humanity would immediately attract the obvious questions: does mainstream evolutionary science accept the theories on which the worldview is based? If not, how can you expect an educated public to take your proposed worldview seriously? Before you seek public support and acceptance, should you not first get the acceptance of the relevant scientific community?

I spent years attempting to do the scientific work myself because evolutionary science had failed to produce the requisite 'big picture' understanding of evolution.

I was primarily interested in evolutionary science because of its ability to answer vital questions about the sense and significance of human existence. I saw its enormous potential to answer fundamental questions about human existence: What are we? In what ways should we evolve ourselves psychologically? How should we live our lives? How should we organise ourselves socially and politically? For what purposes? And so on.

However, evolutionary science had failed to produce the foundations needed to support an evolutionary worldview. Consequently, I had to develop them myself. There was no alternative. I had to go through the laborious process of writing up my evolutionary insights in the arcane form required for scientific papers, replete with scholarly citations of the relevant literature. Considerable time and effort were necessary to dress up my papers so that they looked as if they met all the requirements of academic papers. I have continued this strategy

195

until now, slowly but surely building the scientific support needed to provide the foundations for an evolutionary worldview.

When the evolutionary scientists pulled out of the Evolutionary Salon, I could no longer pursue what had been my main goal. But this freed me to relax and enjoy the show. It was an interesting week, full of interactions with fascinating people. Some of America's leading spiritual and social progressives participated in the Salon.

The Salon did not adopt the standard format in which presenters delivered their prepared papers to an audience. Instead, it utilized less structured processes designed to facilitate collective intelligence and creativity, such as 'Open Space' and 'World Café'.

However, although these processes were effective at encouraging collective participation, they did not tend to produce systematic intellectual engagement. A good time was had by all, but heavy-duty thinking was rarely evoked.

This was not due to any failing of the collective processes that were employed. Instead, it was due to the nature of the majority of the participants.

Speaking somewhat metaphorically, the attendees tended to be 'heart' rather than 'head' orientated; 'right-brain dominant' rather than 'left-brain dominant'; 'feeling' rather than 'thinking' in the Myers-Briggs framework; 'Green' rather than 'Orange' in the terminology of Spiral Dynamics; humanities-orientated rather than science-based; postmodern rather than modernist; and so on. The exodus of the scientists had unbalanced the composition of the Salon.

Consequently, the Salon did not stimulate me to have any new 'big picture' insights into large-scale evolutionary processes and their implication for humanity. Furthermore, even though everyone was very nice and positive about my book Evolution's Arrow, most did not actually understand its central themes. Generally, this was not because of any lack of intellectual capacity. Rather, it was due to a lack of motivation.

Although many participants seemed unaware of this, the ideas they found interesting were those with implications that were uplifting and inspiring to them. For them, Evolution's Arrow had a very inspiring and positive message for humanity. Evolution has a direction, and it is towards increasing cooperation and integration. The history of life on Earth is not about destructive competition, but about increasing integration. How good is that? What more would you need to know? Let's celebrate!

However, the mechanisms that evolution has had to use to overcome barriers to cooperation are not so inspiring and uplifting. As detailed in Evolution's Arrow and earlier in this book, complex cooperation evolves only when powerful managers emerge that use their power to punish free-riders and reward cooperation. When they first emerge, proto-managers tend to predate upon and exploit the proto-organisations that they later come to manage.

These processes have to be understood in detail if one is to apply them to the future evolution of humanity. Such an understanding is essential if humanity is to achieve the next great cooperative evolutionary transition on earth: the emergence of a unified, sustainable, and highly-evolvable global society. In order to accomplish this transition successfully, a deep understanding of the complex evolutionary dynamics that are needed to enable these kinds of transitions is essential. Dispassionate knowledge about how the relevant systems need to be organised and constrained is absolutely necessary. This heavy cognitive lifting might not be inspiring and uplifting to heart-orientated people, but the survival and future evolution of life on Earth depend on it.

Nevertheless, the Salon's right-brained bias did have a useful effect on my evolutionary work. It got me to focus more on how the new emerging evolutionary worldview could be promoted and publicized. This was a major preoccupation of many of the participants. They were very interested in how the power of the worldview could be conveyed through documentaries, movies, computer games, music, and other media.

A particular focus was on approaches that could evoke what we referred to as 'evolutionary epiphanies'. These are intense experiences in which individuals suddenly integrate the various elements of an evolutionary worldview into a coherent whole, and realize in a rush of

197

insight that they, themselves, are important participants in the large-scale evolutionary processes that they are now envisaging. Typically, this produces a shift in identity—the individuals no longer experience themselves as isolated individuals, but as integral and active participants in the evolutionary process itself.

These discussions at the Salon stimulated my next major evolutionary project: I decided to write The Evolutionary Manifesto. It was intended to be the evolutionary counterpart of Karl Marx's Communist Manifesto, only more successful.

The Communist Manifesto was communism's call to action. It identified where human societies were currently deficient and how they needed to be fundamentally re-organised for the benefit of all. The Communist Manifesto appealed not just to the intellect, but also to commonly-held values and emotions. This was consistent with its goal of encouraging revolutions aimed at transforming human societies for the benefit of all. It was designed to motivate powerful action. It was not just a set of ideas, not just candy for the brain.

I was soon given the opportunity to work full time on writing The Evolutionary Manifesto: In June 2007, I gratefully accepted a voluntary retrenchment package from the Department of Industrial Relations. This gave me a lifetime superannuation pension adjusted for cost-of-living increases, the cash value of my accumulated leave entitlements, and a substantial tax-free lump sum severance payment. Given that our daughters were now well-educated, capable, and independent women, my wife decided that these major changes signalled that it was a good time for her to leave.

I had been happily married for 30 years and had just assumed that the marriage would continue until one of us died. But my wife decided that she wanted to come out from under my shadow and paddle her own canoe. Once I had thought about it, I could see the good sense behind what she had decided. If I had been giving her independent advice, I would have recommended that she end the marriage, go out on her own, and run her own show.

Under Australian divorce laws, my ex-wife got half of everything, including half the indexed pension. But my share is enough to enable me to live comfortably for as long as I survive, provided I am frugal and live within my limited means.

At 54, I was single and retired from paid work. I moved to Melbourne and bought a small apartment overlooking Melbourne's Chinatown. I had found during travels in Asia that I loved the energy and dynamism at the center of a large city.

I resolved never to work for money again, even if I was paid full time but only had to work part-time, as before. I received several offers for contract work in the early years following my retirement, and happily turned them all down.

It is now 17 years since I have done any paid work, and counting. For 17 years, I have been a full-time evolutionary activist, pursuing the goal of contributing to the advancement of the evolutionary process on Earth.

In 2008, I completed the Evolutionary Manifesto and a companion document, 'Strategies for Advancing Evolution'. Together, they represented my intentional attempt to promote evolutionary activism and to identify associated strategies. These included strategies for actualizing the next great evolutionary transitions on Earth. Foremost amongst these transitions were the emergence of a unified and highly-evolvable global entity and the development and spread of metasystemic cognition, ultimately enabling an evolutionary awakening and a Second Enlightenment.

I made the Evolutionary Manifesto freely available on its own website (evolutionarymanifesto.com). Together with Strategies for Advancing Evolution, I also published it as a cheap Amazon Kindle book. Hundreds have subscribed to occasional emails about progress in spreading the evolutionary worldview. However, the great majority of these tend to be predominately 'heart-orientated' rather than either 'head-orientated' or a synthesis of both left and right-brain orientations.

In the meantime, I had written a paper on the future evolution of consciousness. It was published in the Journal of Consciousness Studies in 2007. I delivered presentations about the main themes of the paper at the Towards a Science of Consciousness conferences in Tucson in 2008 and in Hong Kong in 2009.

The goal of the paper was to develop a model of brain processes that can explain why meditation-like processes are able to produce a variety of modified states of consciousness that can be useful in particular circumstances. This model was able to explicate the effects of meditation without resorting to the religious and supernatural explanations given by the spiritual and contemplative traditions. The paper uses the model to identify how to improve existing meditation-like processes for particular purposes, and provides the foundations for the development of an array of psycho-technologies optimized for various purposes.[13]

The paper pointed out that this proposed transition in the study of meditation-like practices is analogous to the transitions that have occurred previously in many domains: folk theories and religious explanations of phenomena are replaced by a scientific approach. This, in turn, often drives the explosive development of technologies that make use of the new scientific knowledge.

Ultimately, the successful undertaking of this transition in relation to spiritual experiences and practices has the potential to enable individuals to significantly increase their evolvability: It will enable them to adopt at will whatever state of conscious functioning is optimal for meeting whatever challenges and circumstances they are facing.

The paper gave examples of how its model of conscious functioning could be used to generate new psycho-technologies. For example, the paper used its model to identify how particular states could enhance cognitive processes. In particular, it showed how a state in which consciousness is not embedded in thoughts and feelings could facilitate access to higher cognitive capacities.

In a state in which conscious attention is not contracted down and filled with thinking and feeling, consciousness is able to recruit relevant resources from the unconscious mind. Freed from embeddedness in thought, it can recruit intuitions and also access pattern-recognition capacities. These capacities are essential if cognition is to be able to understand and build models of complex phenomena.

---

[13] Stewart (2007) – see References for full citation

Complex dynamical systems typically include processes that can only be understood and represented in mental models as patterns. Analytical/rational cognition is very limited in its ability to identify and use these complex patterns in its thinking. If complex patterns are to be integrated into cognition and its mental models, it is essential that they can be perceived as a whole. This cannot be achieved by analysis.

This ability to use states of consciousness that can be accessed by meditation-like processes can have the same kind of cognitive benefits as 'sleeping on' a problem or relaxing in a shower. These activities tend to still the mind and disengage the individual from embeddedness in analytical thought and emotion. They facilitate access to intuitions and pattern-recognition capacities from the unconscious mind.

However, meditation-like practices have the advantage that they can train the ability to access these capacities at will, as needed. They enable conscious control and optimization of conscious functioning. Ultimately, they enable the technologizing of consciousness.

The paper also explored how meditation-like technologies could be used to become self-evolving and to see one's thinking and feelings as object.

My participation in the Tucson conference in 2008 provided me with an important new piece of the puzzle for technologizing meditation-like practices. Several workshops were conducted on the day preceding the formal opening of the conference. Always on the lookout for new approaches that I could use to recursively improve my models of meditation-like practices, I attended a workshop presented by Kleo Ormos. She guided participants in a number of powerful meditation practices.

One particular approach that she used to deepen the effects of her practices required the practitioner to concentrate awareness on two or more sensations that are located apart in space. For example, one practice required the practitioner to concentrate on the sensations associated with their finger tips and also with those associated with the tips of their toes, both at the same time. This practice demands that the practitioner concentrate unwaveringly on the two sets of sensations. Doing so quickly stills the mind.

But she also incorporated into her practices an approach that was even more significant and valuable for my purposes. Kleo had specifically designed her practices so that they enable a practitioner to stay anchored in the present, after the formal 'sitting' practice is complete. Her practices were not directed only at producing awakening on the meditation cushion in a quiet and darkened room. Instead, they were intentionally designed to initiate and maintain awakening in the midst of ordinary life.

Once practitioners have entered a meditative state, this design element requires them to rest concentrated attention on a source of sensation deep within their body. When the sitting meditation ends, practitioners begin to move around their environment and start to interact with others. However, all the while they are doing so, they are required to continue to rest part of their attention on those internal sensations. If they find at any time that they have become re-embedded in thought or feeling, they return attention again to the sensations, and so on.

Importantly, when practitioners get up from their sitting meditation, the practice requires them to refrain from moving their full attention away from these internal sensations and onto aspects of their environment or onto thoughts or feelings. The natural tendency that we all have in these circumstances is to do just this. We tend to move our attention back out onto objects and events in the environment (or onto thoughts and feelings), become absorbed in them, and fall asleep again.

But if we use Kleo's method, we must ensure that part of our attention remains anchored in our body. Part of our attention does not shift into the environment and away from the internal anchoring sensations. Instead, when we begin to re-engage with our external environment, we expand our attention out from its concentration on the anchoring sensations, without losing contact with those sensations.

We are to become aware of our environment by expanding our awareness, not by shifting it. Furthermore, as with most meditation practices, if we find at any time that our full attention has moved out into our environment or become embedded in thought or emotion, we move part of it back to the anchoring sensations within our body. And so on, and so on, repeatedly.

I started to incorporate these elements into my awakening practices in the coming weeks. Soon, I began to realize that the awakening capacity that I had discovered while trying to solve Libby's tennis problem, was probably a diversion away from my primary goal. At the time, it seemed like a major advance. It enabled me to awaken to the present, at will. I could immerse myself in vivid, spacious awareness whenever I wanted.

In this state, I tended to experience my environment, wherever I was, as if I were inside a magnificent cathedral (this, of course, was not just a coincidence—cathedrals are intentionally designed to still the mind and produce presence. The propensity of humans to experience vivid, spacious awareness has often been used to convince them of the existence of 'the divine'. But it is a capacity that all humans have, irrespective of whether or not any particular supernatural entity actually exists).

However, being awake in a state of expanded awareness is not easy to maintain. If one's awareness is out in the external environment, it is particularly susceptible to being attracted to outside events. Attention will be very easily distracted. I had developed the ability to come into the present at will, but remaining there was a different matter entirely. I would continually find myself re-embedded in thoughts and feelings about aspects of my environment.

In contrast, if part of one's attention is resting on sensations deep within the body, there is nothing within that attention that is likely to give rise to thoughts, feelings, or other distractions. It is relatively easy to give bare attention to inert, internal sensations and to maintain that attention. Even though the other part of this divided attention is directed outside the body in spacious awareness and is more susceptible to distraction, the part that is resting on inert, internal sensations acts as an anchor.

The more that I practiced returning part of my attention to the anchoring sensations whenever I found that my consciousness had re-embedded, the easier it became to remain anchored in the face of distractions. It strengthened the relevant 'muscles'. This was further facilitated if I surrendered to whatever arose, both internally and externally. Radical acceptance tended to lessen attachment to goals and intentions, further reducing the power of potential distractions.

I continued with these improved meditation and awakening practices. Although I could enter a state of presence at will and was better at maintaining that state, I would still spend long periods during the day in contracted awareness, with attention absorbed in thinking or in talking with others. At least, this was the case until I had an extraordinary experience in September 2008.

As was one of my habits at the time, I was sitting on a bench seat facing the fountain in the small park on the south side of Victoria's Parliament House. I had been reading and sunning myself, but decided it was time for my meditation practice. On this occasion, I had my eyes closed, with my attention concentrated on feelings inside the back of my head. These were the anchoring sensations that seemed to work best for me at the time. In a few minutes, my mind was still and peaceful.

Suddenly, I was disturbed by a loud, grating noise inside my head. I had never experienced anything like this within my head before. Then I realized that the noise was not actually inside my head. In fact, it was the sound of one of Melbourne's trams crossing an intersection about 50 meters behind me. I relaxed again into peace and stillness.

But I began to realize that the sensations associated with all my senses now seemed to arise within me. I maintained the state and opened my eyes. Now, I found that my mind was not distinguishing between objects that are normally experienced as being outside myself, and those that are normally experienced as being located within myself.

My experience no longer appeared to be perceived from a particular point of view that was centered on my body and located at a particular position in space. All locations were now equal. They were not distinguished by where they were positioned relative to my location. Phenomena that arose within experience were no longer perceived from a perspective that originated from where my body was situated.

It is very difficult to describe what I was experiencing. You have to have had such an experience yourself if you are to know precisely what I am pointing to. But if you have had the experience, you will know immediately what I am trying to describe. The experience seemed to me to be what I had read about and heard others talk about as a 'non-dual' awakening. In such an awakening, there is no separation between the 'I' and the external world. All is 'one'. There is just 'the whole'.

I had heard a description of such an experience many years ago that now made sense: the individual reported going to sleep one night as normal. But when he awoke in the morning in his bedroom, he could not tell whether the knobs on the drawers of his dresser were looking at him, or whether he was looking at them.

Adherents of spiritual traditions often refer to such an experience as a 'unitary experience' or an experience of 'oneness with the divine'.

I found that philosopher Thomas Metzinger's book 'Being No One' provided an explanation that is far more simple, plausible, and capable of underpinning a science-based model of the phenomena: He pointed out that for us to be aware of an object or process, there must be a representation of that process in our consciousness. On this basis, if we are conscious of a sense of self that is experienced as separate from other objects in consciousness, then there must be a representation of our self in consciousness. He refers to this representation as a Phenomenal Model of the Self (PMS). On a similar basis, the fact that our conscious experience appears to be organised from a particular point of view is due to the presence in consciousness of a Phenomenal Model of the Intentionality Relation (PMIR).[14]

This simple model can explain the nature of non-dual experience. It suggests that a non-dual experience arises when the PMS and the PMIR are dropped from consciousness. When this occurs, a person's conscious experience no longer includes awareness of the self and of its point of view. Their experience no longer distinguishes between self and other.

In a sense, this involves going a step beyond what is achieved in a standard meditation practice. Typically, the meditator drops all attachment to thought and feeling. Since much of the bandwidth of their

---

[14] Metzinger (2003) – see References for full citation

consciousness is no longer taken up with the conscious processing of thought and feeling, they experience spacious, expanded awareness. Objects in visual awareness, for example, are experienced vividly. Experience is not distracted or contracted down by thoughts or feelings that arise.

After the dropping of thoughts and feelings, the next step in meditation beyond this would be to drop from consciousness representations of the self and its relationships with its environment. In Metzinger's terminology, the meditator drops attachment to their PMS and their PMIR. They enter a 'non-dual' experience.

This simple model of meditation and consciousness can also account for other forms of conscious experience. As a fan of the sport of rugby league, the greatest game of all, I was intrigued to hear one of the great players describe how he experienced himself while playing. He said that as he was running across the field, ball in hand, deciding whether to pass the ball, or continue running with it, or kick it, he experienced himself and the other players from a point of view above his body.

From this vantage point, he could look down and see where the players on the other team were positioned. Immediately, he could see his best option. This ability made him one of the 'all-time great' players. He was not particularly fast or strong, but his consciousness functioned in a way that more than compensated for these limitations.

Other examples include drug-induced states. In effect, particular drugs can prevent the brain from maintaining awareness of a model of the self and a model of the point of view of the self. The dissolution of the model of the self can be experienced as dying. It probably is precisely what an individual might experience in some circumstances if they are actually dying. The dissolution of the internal model of the self is the ending of the feeling of 'I'. It is the end of the experience that there is a doer, including the experience that "I think, therefore I am". Fortunately, however, in the case of psychedelic experiences, the effect is usually temporary.

Initially, the dissolution of the PMS and PMIR produces a non-dual experience. As the effects of the drug diminish, this is followed by a return to normal as the self-models are reinstated in consciousness, and individuals once again experience themselves as a self that is separate from the rest of what is experienced as reality.

Psychedelics can also produce states of consciousness that put together the contents of consciousness from a point of view that is different from normal. It is easy to see how this could be experienced as an 'out of body' state, similar to the example I have given of the rugby league player. In such cases, the difference in point of view is a relatively simple product of how the contents of consciousness are organised. The normal point of view is constructed within the brain, and it can be put together differently, producing a different experience.

These different states of consciousness are potentialities that all humans possess. They can be understood most simply in terms of science-based models. There is nothing intrinsically mystical and non-materialistic about them. However, the world's spiritual and contemplative traditions provide a wide range of spiritual explanations and interpretations of the states. For example, many describe the non-dual state as experiencing unity with the divine. Others suggest that it entails realizing your 'true' self, which is the ground of all and everything. And so on.

Many of the traditions disagree with the interpretations of others. Wars have been fought over differences.

However, a science-based approach strips the supernatural content from the interpretations and explanations of these states. It sets out to understand these experiences in terms of psychological and brain processes that science can recognise. As I have indicated, this has an enormous practical advantage over the stories told by spiritual and religious traditions. It produces models that have the potential to be tested against evidence and used to technologize these capacities.

Models of the brain processes that produce these states can be used to identify practices that are more effective at producing them. We can use the models to try out variations of traditional practices in our heads and in actuality. Bringing these possibilities into the realm and methods of scientific inquiry enables their full potential to be explored systematically.

The spiritual traditions have passed on their practices successfully over many generations. This is no small achievement given that complex practices are likely to be degraded and modified unrecognisably as they pass through many hands and minds. However, this has been achieved at the cost of freezing the practices and preventing innovation and creativity. To ensure that the essential elements of the practices are preserved over the generations, they have been embedded in immutable traditions. The practices and their explanations became sacred, and were therefore protected from experimentation and tinkering.

In order to survive, the practices that were frozen by tradition also had to be in forms that posed no threat to powerful authorities. Somewhat ironically, it was essential that they be preserved in forms that interest me the least, and not in forms that are of critical importance for the successful future evolution of humanity. The practices that were handed down unchanged over many generations were of little practical use in real-world activities. If the practices had been in a form that, for example, could be used to enhance effective agency in the world, they would have threatened the interests of the powerful. In order to protect their interests, the powerful would have tried to appropriate the practices to strengthen their power and to exclude their enemies from accessing them.

In recent human history, those who were powerful in a particular region at a particular time did not last long. When rulers were inevitably overthrown, any practices and traditions that they sponsored were likely to be destroyed along with them, particularly if they were seen as a threat to the new rulers. In any given place on the surface of the planet, the powerful have been overthrown many, many times during the last 5,000 years of human history and evolution.

Arguably, if practices and their justifications were to survive throughout human history until the present, they had to be treated as immutably sacred, and also be in a form that could not enhance agency, particularly in war-like activities. Ideally, they needed to be preserved in isolated communities that were of no threat, such as in monasteries in unproductive, rugged, mountainous regions. Or in small, secret groups within larger societies that were perceived as benign by the powerful (e.g., mystery schools). Right-hand path traditions that promoted surrender to the absolute rather than enhanced agency, were pre-adapted to survive.

Gurdjieff suggested that the allegory of Noah's Ark contains a deeper message about the need to preserve spiritual practices and knowledge during 'dark ages' characterised by incessant warfare. The Ark is a metaphor for a small, isolated spiritual community that preserves what is valuable from destruction amidst the deluge—the continual war of all against all.

The fact that practices have survived does not mean that it is because the practices and the beliefs that supported them are justified and valid. It is because they had characteristics that facilitated their survival.

But the circumstances that stood in the way of using scientific methods to further develop and technologize practices that enhance agency in the world, apply no longer. The monasteries and mystery traditions have done their job.

It is now safe to exploit the enormous potential of meditation-like practices to enhance human functioning.

The non-dual state that I experienced in 2008 did not come and go during the day. It seemed permanent. It was there before I went to sleep, and it was there when I awoke in the mornings.

Abiding in nan-dual awareness was an extremely pleasant and peaceful state. Things just seemed to happen. I did not have to get involved in them or figure out what to do. Things seemed to work out well despite the absence of worry and protracted thinking.

In retrospect, this was not surprising: my body-mind had accumulated a large store of learned behaviours that it could draw upon to have me behave appropriately in most situations. I did not have to get involved in deliberating about what I should do. I could rely on my prior conditioning and learned behaviours to initiate actions that were apt. This reminded me of an oft-repeated statement made by teachers in the School for Self Knowledge: if you are present, surrendered, and non-attached, you will spontaneously act in ways that are appropriate to the needs of the situation.

For a week or so, I wandered happily around the streets of Melbourne in this state. I described it as being in 'la la' land. I was not permanently in the state of bliss that is often described by some spiritual traditions as accompanying awakening and enlightenment. But if I stayed in the state for the rest of my life, it would be a life of happiness and ease.

If I was pursuing the right-hand path, I would have surrendered completely into the state for the rest of my life. There was nothing that I would have wanted to do in the world. And the non-dual state was perfect for not doing.

But I was on the left-hand path. There were things in the world that I wanted to do. I was firmly into acting in the world. I had evolutionary goals, and I wanted to develop my capacities to achieve those goals.

Happiness had never been my central objective. For individuals like me who were fortunate to be born in the 1950s in Australia, it was not a challenge at all to arrange your life so that you were happy. Free university education, plenty of well-paid jobs, cheap housing, no wars within Australia, steadily increasing economic prosperity, and so on. But from an evolutionary perspective, a happy life is not necessarily a good one. It will not always lead to survival and flourishing.

So after about a week and a half, I decided to bring my period of persistent non-dual awakening to an end. But first, I wanted to test whether I could return to 'normal' and then get back into the non-dual state using the technique that I thought had taken me there originally. After returning to normal for a day or so, I began concentrating my attention on feelings at the back of my head, while surrendering and radially accepting all that arose in expanded awareness. After a couple of hours, I was back in persistent non-dual awakening.

I stayed in that state for a few days until it was time to go to the annual conference of the Jean Gebser Society that was being held that year in Melbourne. In the 20th century, Gebser was one of the great explorers of the evolution of consciousness. Somewhat ironically, my

210

wish to participate in a conference about his ideas prompted me to abandon my exploration of persistent 'higher' levels of consciousness, at least for the time being.

This experience played an important role in shaping my understanding of the evolutionary significance of the various states of consciousness that humans can enter. It is clear that humans have the potential to experience a variety of states of consciousness. Some of these states can enhance agency and adaptability in particular circumstances, while other states do so in different situations. Given my evolutionary goals, I had a particular interest in accessing states of consciousness that facilitated the development of higher levels of cognition, as well as a capacity for self-evolution.

From the point of view of right-hand paths, surrendering into permanent non-dual awareness might be considered to be the ultimate goal—absorption into the absolute. But in my experience, it was not a useful state for many other purposes. In particular, I did not find it advantageous when I needed to achieve goals that required deep cognitive work and effective agency in the world.

I reached the conclusion that evolvability would be maximized by developing the ability to enter whatever state of consciousness was optimal for achieving one's goals in whatever circumstances one faced. No individual state was beneficial in all circumstances. It was a matter of horses for courses.

Consistent with this conclusion, I accepted that the non-dual state could be very useful in some circumstances. In particular, I decided that if I ever ended up decrepit, demented, and incontinent in a nursing home bed, with no hope of recovery, I would attempt to use my techniques to enter a persistent non-dual state. I would do the same if I was going to be tortured. This raised in my mind the iconic image of the Buddhist monk dousing himself with petrol on the steps of the American embassy in Saigon in the 1960s, setting himself on fire, and burning unflinchingly.

Fortunately, I have not yet ended up in a nursing home or been tortured. But somewhat unfortunately, a decade or so later, I had an opportunity to put these strategies to a challenging test.

211

HUMAN SUPERINTELLIGENCE

17.

A Breakthrough for Scaffolding Metasystemic Cognition

In 2009, I read a book review by Sara Ross in the Integral Leadership Review that drew my attention to an entirely new pathway for advancing the evolutionary process. Up until that point, the recursive improvement of my cognition and my drive to understand large-scale evolutionary processes and other complex phenomena had enabled me to develop metasystemic cognition. My evolutionary and psychological publications evidenced this.

But at this time, my knowledge of metasystemic cognition was largely procedural, not declarative.

As I mentioned earlier, procedural knowledge is the knowledge that enables us to exercise a skill successfully. It is knowledge of how to do things. It is generally unconscious, and we are unable to put procedural knowledge into words. If you have a good serve in tennis, you will likely be unaware of the actual muscle contractions and bodily movements that constitute your service action.

In contrast, declarative knowledge is the conscious knowledge of facts, propositions, and analysable models. It is knowledge about things, and can be described in words. This enables declarative knowledge to be conveyed to others in writing or other forms of language. It can be taught explicitly.

I had developed a comprehensive set of metasystemic cognitive skills that worked effectively for me. However, I did not have detailed declarative knowledge about those skills and what my metasystemic thinking entailed. Consequently, I could not explain comprehensively to others how to do what I did.

I was like champion tennis players who have a powerful and accurate serve but do not have a declarative model of the movements and actions they use to produce it.

Through persistent practice and repetition, individuals can learn procedural knowledge that enables them to exercise skills with great ability. But this does not necessarily enable them to impart this knowledge to anyone else. Nor does it enable them to understand how their skill can be improved or adapted to novel circumstances. They generally need a declarative model for this. In part, this is why it is common in sports for a great player to be a poor coach.

I knew that this lack of declarative knowledge about metasystemic cognition was a major impediment to achieving my evolutionary goals. To recap, my evolutionary theorizing demonstrated that the next great step in cognitive evolution in humans is the development of metasystemic cognition. First Enlightenment thinking and analytical/rational cognition began to spread widely a few hundred years ago, and powered the rise of science and technological development. However, this thinking is incapable of generating effective mental models of complex phenomena, including large-scale evolutionary processes. Unfortunately, most of reality is too complex to be understood by analytical/rational cognition. Science has not yet developed the capacity to understand and model complexity.

This is one of the most serious challenges facing humanity: the existential threats that have the potential to end human civilization this century are too complex to be understood and overcome by analytical/rational cognition. The transition to metasystemic cognition is essential for understanding and embracing an evolutionary worldview and for enabling the survival and flourishing of humanity.

Sara Ross' article reviewed Otto Laske's 'Manual of Dialectical Thought Forms'.[15] Just reading the review itself got me extremely excited. It was clear from the outset that the Manual contained the declarative knowledge that I was hoping either to find or to develop myself. The Manual identified 28 'thought forms' or 'movements in

---

[15] Ross (2009) – see References for full citation

thought' that thinkers need to use to build effective models of complex phenomena.

The thought forms were grouped by the Manual into four classes or Quadrants: those dealing with the context of the phenomena that are being modelled; those dealing with the processes that constitute the phenomena; those dealing with the evolving relationships between the processes; and those that deal with the phenomena as 'systems-in-transformation'.

These four categories of thought forms identify what is typically left out of the models developed by analytical/rational thinkers. Their models tend to focus down on some limited and relatively isolated phenomena that they seek to understand. As such, their models tend to ignore the context in which these limited phenomena are embedded. This approach works well in simple physical phenomena where laws apply irrespective of the specific nature of the context in which the simple phenomena are situated. However, this kind of approach often fails to model complex systems where the context is highly relevant to how the phenomena unfold.

The mental models of analytical/rational thinkers also tend to ignore the fact that all objects/things in the world are actually processes, not objects with fixed attributes. On a time scale that is sufficiently long, even mountain ranges are processes, not objects with fixed attributes. Everything in reality evolves and changes ceaselessly.

The third class of thought forms that tend to be ignored by analytical/rational thinkers are those that recognise that processes that constitute complex phenomena are often related to one another, and tend to coevolve through time as the phenomena unfold dynamically.

Finally, the 'systems-in-transformation' thought forms recognise that when the context, process, and relationship thought forms are put together appropriately, the resultant models can capture the reality that complex phenomena generally constitute systems in transformation as they evolve and coevolve.

In these ways, the four Quadrants of thought forms identify what tends to be left out by analytical/rational thinking. The mental models generated by analytical/rational thinkers tend instead to produce mechanistic models. Analytical/rational thinkers build models that are generally constituted by objects (or sometimes by simple processes) that have fixed attributes. These attributes tend to be unchanged as the objects or simple processes interact according to fixed rules or laws.

Analytical/rational models can only build and mentally simulate models that are analysable, and that therefore can be 'thought through'. The paradigmatic example is the Newtonian model of particles of matter that interact like billiard balls, in circumstances where the context can be ignored and the particles do not change their attributes endogenously as processes, or as a result of interactions between them.

One way of experiencing the fundamental differences between analytical/rational thinking and metasystemic cognition is to notice the kind of world that is built by analytical/rational thinking, and to compare it with the complexity of most of the rest of reality.

The built environment in which humans live has been conceived and designed by analytical/rational thinking. Consequently, it is comprised of structures and processes that are capable of being analysed and thought through. Humans have little ability to incorporate complex phenomena into their built environment. Whatever we build must be understood by analytical/rational cognition and be intelligible to it.

With this in mind, consider the center of any large modern city. The structures you see will commonly exhibit large, flat surfaces that meet at right angles. Box-like shapes and straight lines characterise almost everything you can see. Motor vehicles are machines in which parts with fixed attributes interact in predetermined ways to cause the vehicle to fulfill designed functions.

Now compare this with what you experience when you walk through a mature forest. Right angles, boxes, and flat surfaces are extremely rare. Any organisms that you see are organically organised, not machine-like. Almost anywhere you look closely, you see a buzzing confusion of interaction at various scales of space and time. Using analytical/rational cognition, you will be unable to think through the ceaseless change going on around you, and will be unable to predict how it will unfold through time.

216

Most of the reality you encounter in a mature forest cannot be understood using mechanistic mental models. More generally, you will find that most phenomena in the world cannot be understood adequately and modelled by analytical/rational thinking or by the science that it powers.

By identifying the thought forms that tend to be missing from analytical/rational thinking, the Manual of Dialectical Thought Forms provides an invaluable aid for scaffolding metasystemic cognition in analytical/rational thinkers. These missing thought forms identify the additional aspects that the analytical/rational thinker needs to include in their mental modelling if it is to represent complex phenomena adequately. The relevant thought forms identify the aspects of phenomena that the thinker must give attention to if they are to understand complexity.

To build a capacity for metasystemic cognition, an analytical/rational thinker needs to practice making use of the thought forms to identify where their mechanistic models are deficient. Then, they need to move their attention to these absences in their models and to remedy what is deficient. More specifically, they need to consider whether any relevant context needs to be included in their modelling; whether particular objects within their models might be better-represented as processes; whether it is useful to include co-evolving relationships in the models; and whether it is useful to model the phenomena as systems-in-transformation.

Of course, this does not mean that a thinker needs to use each and every thought form to build each and every model. As a general rule of thumb, models should be as simple as possible, given their purposes, but not more simple. Consequently, a metasystemic thinker will continue to use simple analytical/rational models where the phenomena they are interested in can be approximated adequately by such a model, given the goals of their modelling.

Individuals can use Laske's declarative knowledge about model building to help scaffold metasystemic cognition in themselves. Or coaches can use it to scaffold metasystemic cognition in others. At first, this is a slow and laborious process in which the individual learns to use particular thought forms to identify specific absences in their mental modelling, and then attempts to use the relevant thought forms to remedy each specific deficiency.

However, with persistent effort and practice, the individual will be able to implement the new movements in thought fluidly and unconsciously. Through practice, they will have converted the declarative knowledge into procedural knowledge.

This is comparable to the way a developing tennis player receives declarative knowledge from a coach about how to improve their serve. But in order to become proficient in the new service action, the player must then spend considerable time practicing, receive more coaching, and so on, recursively, until the new service action can be performed fluently and unconsciously.

As mentioned earlier, modern complexity science is almost entirely a product of analytical/rational thinking. It is the outcome of attempts to apply mechanistic analytical/rational thinking to complex phenomena that cannot be modelled adequately by it. As analytical/rational thinking spread during the First Enlightenment, it achieved early success in the modelling of simple physical phenomena such as forces, motion, and gravity. This led to attempts to use it more widely. These attempts were successful when they were applied to aspects of reality that happened to be able to be understood by analytical/rational models, e.g., because they were relatively mechanistic. However, they failed when phenomena were too dynamically complex to be modelled adequately by analytical/rational thinking.

This was not a problem for science at the time. Scientists who operate at the analytical/rational level are generally not aware of what their models are missing. The aspects of complex phenomena that are absent from their models obviously cannot be identified by their thinking—these aspects are missing from their thinking. Scientists cannot think about or understand what they cannot think about.

Scientists who view reality through the prism of analytical/rational thinking will not see the limitations of their thinking. Instead, they will conclude that their analytical/rational thinking potentially enables them to understand everything that can be thought about effectively. They tend to believe that First Enlightenment science can answer all questions that can be answered adequately.

Some of the finest minds of the 20th century began to see the limitations of analytical/rational thinking in relation to complex phenomena. Several of them led attempts to build a new science of complexity. However, none of these attempts made significant progress. Primarily, they failed because their complex discoveries would be accepted by mainstream science only if they met the narrow, mechanistic criteria used by First Enlightenment science. Whenever it looked like progress was being made, they would be pulled back in again by the mainstream. This has continued to restrict mainstream science to the exploration of phenomena that can be modelled effectively using only analytical/rational thinking.

As a further consequence, what currently passes for complexity science is often no more than an analytical/rational reduction of complex phenomena. It has succeeded in understanding complexity only in relation to those limited aspects of complex phenomena that happen to be able to be modelled at least approximately by mechanistic, analytical/rational thinking. As such, up until the present, complexity science has fallen far short of producing adequate understanding of complex social, economic, political, psychological, and ecological phenomena. This failure will continue until a Second Enlightenment drives the spread of metasystemic cognition, and produces Second Enlightenment science. This science will transcend but include existing science, and will have its own criteria for assessing whether research and theorizing constitutes proper science.

I was particularly excited by Otto Laske's Manual of Dialectical Thought Forms because it was not a weak, first attempt to identify the movements in thought that were needed to constitute metasystemic cognition. It was highly detailed and insightful about each thought form. It was comprehensive and masterful. For many years, I had been on the lookout for declarative knowledge about metasystemic cognition that could be used to scaffold it. In all those years, I had not discovered anything at all of value. It was a shock to suddenly come across something fully formed and highly developed like the Manual. As can happen with great works of art, it was as if it had just appeared out of the ether, fully developed and without being preceded by a long sequence of less effective antecedents.

It was even more surprising considering that a similar exercise has not yet been conducted for analytical/rational thinking itself. Modern education systems do not scaffold analytical/rational thinking directly in school students. There is no Manual of Analytical/Rational Thought Forms that can be used for this purpose. In significant part, this is because in order to see one's thought processes as objects in awareness, and therefore to be able to describe them declaratively, one has to be at a meta-level of awareness to those thought processes. Only about 30 percent of students in modern societies now reach the analytical/rational level. Very few reach a level that is meta to that.

I was to discover later that the idea of identifying movements in thought came first from developmental psychologist Michael Basseches. He conducted interviews with individuals who seemed capable of complex cognition, and recorded their movements in thought. His method of investigating thought forms by interviewing others does not require that the researcher be at a meta-level to their own cognition. It is much easier to see the movements of thoughts in others, than to see one's own thought processes in real-time.

Subsequently, Otto Laske built on Basseches' work to produce the Manual. But Otto, who had trained in developmental psychology at Harvard, was able to go considerably beyond the work of Basseches. Earlier in life, he had also studied at the Frankfurt School under Theodor Adorno, a renowned German philosopher and dialectical thinker. It was with Adorno that Otto was trained in dialectical thinking, and gradually developed the crucial ability to see his own thinking as object. This unique combination of experience and training enabled Otto to produce the Manual. Its originality and comprehensiveness could have resulted only from such a novel synthesis of different perspectives. Furthermore, it could only have been produced by a person who had developed to a meta-level in relation to their thinking.

Given my long interest in techniques for recursively improving cognition and my awareness of the evolutionary significance of spreading metasystemic cognition, I knew what to do with the Manual when I found it. At last, there was a tool that could be used to scaffold metasystemic cognition in others.

Initially, I used the Quadrants and thought forms on myself and on my own cognition. I set out to test my own model-building capacities against the thought form framework. Did I use all the thought forms in all four Quadrants that were relevant when building mental models of particular complex phenomena? Were there blind spots that I needed to remedy? If so, how could I go about incorporating all thought forms into my thinking so that their use becomes fluid and automatic?

In general, I found that my model-building skills tended to use all the thought forms and more. This was not surprising given that 'big picture' evolutionary thinking is a paradigmatic example of metasystemic thinking. You cannot get far without adequately representing relevant contexts, processes, co-evolving relationships, and systems-in-transformation.

As part of this process of self-examination, I set out to label and interpret my thought processes using the thought form framework. In effect, this involved what is known as procedural to declarative redescription. This process is perhaps best understood by taking a familiar example.

As they develop and learn, infants acquire procedural knowledge in the form of various skills. Initially, they are unable to describe in words the knowledge that underpins these skills. However, as they continue to grow and develop, they learn to translate at least some of the procedural knowledge that they have acquired into declarative knowledge.

This is the process known as procedural to declarative redescription. For example, many children can speak reasonably well by age five. Their speech follows many of the rules of grammar. However, they have no declarative knowledge of the grammatical rules that apply to their speech. During their schooling, some of this procedural knowledge is redescribed into declarative knowledge.

In many domains, the redescription of procedural knowledge into declarative knowledge enables the development of science-based models of procedural skills. This, in turn, enables skills to be generalised and adapted quickly to new circumstances. Knowledge is no longer limited to specific situations.

Consciously and intentionally, I went through such a redescription process in relation to my metasystemic thinking. As I proceeded, it enabled me increasingly to see my thought processes as exemplars of a declarative model, and to talk to myself and to others about my thinking. This made it easier to see my own movements in thought as object, and to consider whether alternative thought forms might be more useful for the building of a particular model.

Using the thought form framework in these ways enabled me to see how the framework could be used to spread metasystemic thinking more widely. I now understood how others could be given the cognitive tools needed to self-scaffold metasystemic cognition. I could also see how others could be coached in their use of the thought forms in order to get them to the critically important stage where they could then self-scaffold metasystemic cognition. A set of tools was now available to catalyse the spread of metasystemic cognition and to kick-start the Second Enlightenment.

However, coaching metasystemic cognition was not something that I wanted to spend the rest of my life doing. My central interests and capabilities are to contribute to the application of metasystemic cognition to understand the future evolution of humanity and to guide it. I am a loner and an introvert, and I was not attracted to spending my days helping individuals or small groups to develop complex skills.

So, I needed to inspire others who possess suitable personality attributes and skills to begin the detailed work needed to scaffold and spread metasystemic cognition.

Fortunately, our market-based economic systems can be relied upon to drive the spread of metasystemic cognition, once it becomes clear that it can be scaffolded in others and provides significant competitive advantages to businesses whose executives have developed it. Competition between businesses drives the spread of any innovation that can provide an advantage to those who adopt it. Businesses without the innovation will be outcompeted by those that do. All businesses in a particular market will have to acquire any significant innovation, or perish. Economic markets provide a powerful incentive for such an innovation to spread.

Metasystemic cognition can provide such a significant advantage. Modern corporations tend to be embedded in highly complex environments. These include competitive markets, systems of governance, political systems, research programs, wider economic systems, and so on. To navigate these complex systems, corporations need models that enable them to predict how these interacting systems will unfold. Any corporation that is better at anticipating the complex consequences of its actions will have a clear competitive advantage.

As we have seen, analytical/rational cognition is not up to the task of producing effective models of the complex challenges faced by modern corporations. Mechanistic, analysable models cannot comprehend the dynamical complexity encountered by corporations. But hardly any executives of major corporations possess metasystemic cognition to a significant degree. Some are extremely intelligent, but they operate largely at the analytical/rational level.

As a consequence, the executives of modern corporations tend to be in over their heads. They are incapable of adequately meeting the complex demands of their jobs. Often, they are faking it. This is one of the key reasons why a must-have capacity for senior executives in large corporations is the ability to act with confidence and self-belief in their decision-making, despite the overwhelming evidence that they do not know what they are doing.

In this context, any corporation that can find a way to elevate its executives' cognitive capacity to the metasystemic level will achieve a major competitive advantage. Its competitors will have to follow suit.

Furthermore, any evolutionary activist who wants to accelerate the transition to metasystemic cognition among humans needs to take advantage of this economic dynamic. They need to contribute to the development and packaging of tools to scaffold metasystemic cognition in forms that will tend to be spread rapidly by market forces. When packaged in such forms, the spread of metasystemic cognition will be autocatalytic.

In fact, it was a similar kind of economic dynamic that drove the emergence of the First Enlightenment and the spread of analytical/rational thinking. Before the rise of mercantilism and large-scale markets, feudalism was the predominant form of social organisation in much of Europe. Feudalism was stable for many centuries. A significant cause of this stability was that innovation and creativity were actively suppressed in feudal systems. Tradition was the prime determinant of how things should be done, century after century. Anyone who challenged traditional religious practices or farming techniques was quickly censured. Members of feudal societies typically felt threatened by change. Safety and certainty were to be found in the status quo.

The rise of mercantilism disrupted this status quo. In the beginning, many members of feudal systems opposed mercantilism strongly. But it was like an acid that persistently eroded the pre-existing social order.

Before the First Enlightenment, analytical/rational thinking was rare. It had flowered in some limited areas at particular times in history. Ancient Greece is an obvious example. But even then, there may have been only a few hundred people at any time who were capable of analytical/rational cognition or higher.

But eventually, the rise of mercantilism in Europe strongly incentivized the acquisition of analytical/rational thinking. Under feudal systems, concrete operations thinking was widespread. It was sufficient to enable individuals to function effectively in social systems where knowledge of traditions and associated skills were all that was needed.

However, merchants could be much more successful if they were capable of analytical/rational thinking. This ability enabled individuals to model the future consequences of their decisions, provided the circumstances were not too complex. A merchant who could plan ahead to some degree could easily outcompete other merchants who were equipped only with concrete operations thinking. Individuals at the concrete operations level could not think abstractly, and thus their ability to generalise their thinking was limited. Analytical/rational merchants could see opportunities and possibilities that their concrete operations competitors were blind to.

Strong incentives to acquire analytical/rational thinking still exist in current societies. We now live in a world where analytical/rational thinking is almost essential if one is to earn a high income. This has led to the creation of compulsory education systems in modern societies. These systems are attempts to institutionalise at a societal level the goal of scaffolding all citizens to at least the level of analytical/rational thinking.

However, as I have mentioned, the limited research that is available reveals that only about 30 percent of the population in Western democracies achieve this level of thinking capacity. In large part, the failure of our current educational practices is due to the absence of sufficient individuals at the metasystemic level. The challenge of scaffolding individuals to the analytical/rational level is too complex to be understood by analytical/rational cognition alone. As we have discussed, it is only with metasystemic cognition that individuals are able to model and manipulate complex phenomena, including complex psychological and educational systems.

These understandings led me to recognize a key difference between the First Enlightenment and the coming Second Enlightenment. The First Enlightenment does not appear to have been conceived, designed, and implemented intentionally by humans, individually or collectively. Fundamentally, this is because to do this effectively requires metasystemic cognition. In order to plan and catalyse the First Enlightenment systematically, the capacity to construct mental models of complex interacting systems would have been essential. But at the time of the emergence of the First Enlightenment, this capacity was almost non-existent, and remains so even today.

I am not suggesting that there were no individuals during the First Enlightenment who knew broadly what was going on with the emergence and spread of analytical/rational thinking. At times, some individuals and groups also took steps to promote the spread of ideas that were produced by analytical/rational thinking. The Freemasons and the founding fathers of the United States of America are prime candidates for this kind of possibility. But there is no evidence of the systematic planning and implementation of comprehensive interventions designed to scaffold and spread analytical/rational thinking itself, beyond the kind of educational activities we see today.

These kinds of considerations led me to the conclusion that the Second Enlightenment should be planned and driven intentionally, at least until economic forces take over.

As activists are wont to say: "If not now, when? And if not you, who?"

So, I teamed up with an American, Lawrence Wollersheim, to hold what we called 'The First Planning Meeting for the Second Enlightenment.'

The stated purpose of the meeting was to develop strategies and techniques to produce a global shift to metasystemic/dialectical cognition. The key issues to be addressed by the meeting were: What new ways of thinking will enable humanity to understand and manage the complex challenges that we are facing? Is it possible to train people in these new ways of thinking? If it is, what strategies will spread the higher forms of thinking to sufficient people across the planet in time to make a real difference before global crises undermine human civilization?

Lawrence lived on a yacht at Sausalito in San Francisco Bay. The yacht was large enough to host fifteen or so people for the Planning Meeting.

But who else should we invite? Ideally, the participants in the Meeting would need to have two capacities that, unfortunately, were then extremely rare, and still are.

First, they had to possess the ability to function cognitively at the metasystemic level. Without this, they would be unable to develop and operate the complex mental models needed to identify strategies designed to spread metasystemic cognition across the planet. Furthermore, a capacity for metasystemic cognition is a prerequisite for developing practices and approaches that can scaffold metasystemic cognition in others. If individuals are to be able to develop such approaches, they must be able to construct and use mental models of complex psychological processes.

The second characteristic that prospective participants needed to have was a history of working on themselves. Participants with some experience in the intentional use of meditation and other practices to transform themselves were critically important for the success of the Meeting. This was because the spread of metasystemic cognition would necessitate the scaffolding in others of a capacity to self-scaffold higher cognition.

Individuals could be introduced to metasystemic cognition by suitable coaching and training approaches. Ultimately, however, if they were to develop a comprehensive capacity, they would need to develop the motivation and skills to self-scaffold. Unless participants had persistently worked on themselves to enhance their own cognitive and psychological functioning, they would not have the experiential knowledge to understand the scaffolding process, and how it might be developed in others.

It is worth noting here that simply having an extensive meditation practice alone was nowhere near sufficient for the purposes of the Meeting. The great majority of meditators in the Western World today do not meditate in order to build their cognitive capacities or to become self-evolving amid ordinary life. Instead, many meditate to achieve states that make themselves feel good. For example, they may meditate primarily to reduce stress, produce internal peace, attain 'higher spiritual states', make themselves feel calmer, and so on.

Otto Laske embodied these characteristics and was the first person we invited. He was enthusiastic about participating, and the First Planning Meeting for the Second Enlightenment was up and running.

However, it was challenging to find others with the desired combination of capacities. Fully developed metasystemic cognition is very rare in mainstream academia, including in the sciences. Analytical/rational thinking predominates. Furthermore, individuals who have persistently worked on themselves to enhance their own cognitive capacities are even rarer, both inside academia and without.

For these reasons, we decided to invite several leaders from the integral community. This community had formed around the work of American philosopher and writer Ken Wilber. His work aimed to produce a 'big picture' framework for all human knowledge. Wilber's writing had a particular focus on spiritual development, but he had become widely known for his ability to produce novel syntheses of knowledge from diverse domains. A number of highly-talented individuals with various backgrounds had emerged around Wilber and his work and were independently building upon it.

Although they might not have fully met the participation criteria, they were very smart and capable individuals who had broken the shackles of analytical/rational thinking to some extent. Furthermore, as far as we knew, they each had a long history of using meditation-like practices, at least recreationally.

The Meeting took place in September 2011. Everyone seemed to enjoy and be stimulated by the discussions. However, the meeting did not produce any breakthroughs or silver bullets for spreading metasystemic cognition. Furthermore, it became evident that few of the participants had much experience in using meditation-like practices or other methods for self-scaffolding higher cognition.

Nevertheless, having to plan and conduct the meeting stimulated Lawrence and me to do more work on our own ideas about metasystemic cognition and how it might be scaffolded and spread. It caused us to examine our own cognitive processes even more deeply and to test against our own experience the suggestions that were advanced by participants. Our involvement enhanced our ability to see the operation of metasystemic cognition in our own minds, assisting us further in observing our thinking from a meta-level.

I had hoped for more, but the main benefit of the meeting was to serve as a symbol for future potentialities and possibilities. The First Planning Meeting for the Second Enlightenment had taken place. The Meeting had brought into concrete reality the notion that the Second Enlightenment differed from the First in that it would be planned and spread intentionally and consciously. It is not possible to predict what specific beneficial effects might be produced by the Meeting in the future. But it serves as a permanent attractor for actions and strategies directed at actualizing a Second Enlightenment. It increases the likelihood that Meetings with similar goals will arise in the future.

The obvious next step in kick-starting the Second Enlightenment was to develop a detailed program to scaffold metasystemic cognition in key target groups, including senior executives in business and government. However, it was evident from the Planning Meeting that none of the other attendees would do this anytime soon. They had neither the motivation nor the capability. If it were to happen quickly, I would have to do it myself, despite my aversion to spending much time in social interactions.

On my return to Australia, I met with my colleague Victoria Wilding to design such a program. She had long been involved in implementing developmental practices. We had previously discussed in broad terms the kinds of approaches that could be used to scaffold metasystemic cognition.

Eventually, we produced a program of ten weekly sessions. The overall goal of the program was to enhance the cognitive capacity of participants in order to enable them to strategize and act effectively in complex circumstances. Importantly, this was not limited to developing metasystemic cognition. As I have mentioned previously, two capacities are required if one is to respond effectively when faced with novel complex challenges. The first is the cognitive capacity to build mental models of the complex challenges and to use these to identify effective strategies.

The second is the ability to actually implement those strategies in the face of emotional and other psychological predispositions that conflict with the actions they need to take to implement the strategies. As I have mentioned, they need to be able to align their pre-existing psychological goals and motivations with their new strategic goals.

Consistent with this overall goal, the first half of the program was directed at developing the capacity to access presence at will in the midst of ordinary life. The purpose of this was three-fold: first, to enable participants to see their thought processes as object so that they would be able to identify where their existing cognitive processes are limited, and how they might be enhanced; second, to enable participants to access intuition, pattern-recognition capabilities, and other cognitive resources that are essential for building mental models of complex phenomena (access to these 'right-brain' resources is usually blocked by continual embeddedness in incessant thinking); and third, to free participants from the dictates of genetic, cultural and social predispositions (including their conditioning) that otherwise causes them to behave in ways that conflict with their goals (individuals who can remain in the present in calm, spacious awareness are able to, for example, free themselves from negative emotions by letting them arise and 'pass through them', rather than being embedded in them and being controlled by them).

The second part of the program was directed specifically at the development of metasystemic cognition. Participants would be introduced to the use of Laske's Thought Form Framework through relevant exercises and practices. They would be required to use the skills developed in the first half of the program to become aware of their thinking and to identify where their existing thought processes were inadequate for understanding complex phenomena. The participants would then be guided to use the Framework to identify how their pre-existing cognitive processes needed to be modified so that they could model complex phenomena more effectively.

The program focused more on doing rather than on transmitting theory. Exercises and practices undertaken as homework were a key feature of the approach. The outcomes of these were then the subject of group discussion and learning.

The next step for Victoria and I was to undertake a pilot of the program. We assembled a group of 10 participants. In general, they were all people who were associated in some way or other with Melbourne's progressive community. Many had been influenced by Ken Wilber's writing, and a few, but far from all, had an established and regular meditation practice.

In general, the participants were enthusiastic, intelligent, and very open to new ideas.

230

The ten-week program went well. The formal evaluations we received from each of the participants at the completion of the program were all overwhelmingly positive.

However, I was unsatisfied with the outcome. I felt that the group discussions about practices revealed that the strategies embodied in the program were unlikely to transform the participants deeply, at least in the short term.

Ultimately, the program's goal would be met only if it produced graduates who were committed to self-scaffolding higher capacities throughout the remainder of their lives. Ten weeks alone was nowhere near enough to produce an individual who was self-evolving and fully equipped with metasystemic cognition. The full development of these capacities required years of highly motivated, recursive work on oneself. In this context, what we had hoped the program would achieve was to give graduates the tools and motivation to embark on this path and to continue on it indefinitely.

I felt that the program did not achieve this. However, what it did do is open to all participants the possibility that they could embark on serious self-scaffolding at any time in the future. As a result of their participation in the program, they knew that this opportunity was open to them, and they knew the kinds of tools they could use if they took it up. There is an old saying in self-development circles that "When the pupil is ready, the teacher will appear." A variant of this insight is that when the pupil is ready, the significance of the teaching they received years earlier will be understood fully for the first time, and they will be motivated to do the work necessary to embody it.

I learned several important lessons from running the Pilot. First and foremost, I realized that the program needed to place an even greater emphasis on getting each participant to practice self-scaffolding. Furthermore, it was important that this self-scaffolding not be limited to scaffolding the building of better models. It also needed to include the repeated and recursive practicing of the self-scaffolding of meta-models, i.e., of models that are used to build new models and to improve them.

The strategy that I developed subsequently to achieve this was to get individuals to identify a complex challenge that they were already

231

highly motivated to solve. Then, without reference to existing sources of knowledge about how to deal with this specific challenge, individuals would be required to build relevant models from first principles. These would include models of the relevant problem and of possible strategic responses. Individuals would be guided to use the Thought Form Framework to assist their model building. Importantly, they would also be asked to give particular attention to the meta-strategies they used to build these models.

When they completed this first phase of model building, individuals would then test the models and strategies against some existing sources of relevant knowledge. If they found that these existing approaches were superior in some ways to theirs, they would be asked to do two things: The first would be to amend their models as necessary in order to incorporate the improvements.

The second would be to identify why their meta-models had produced flawed models in the first place. For example, they may find that their meta-models failed to use particular thought forms or interpreted relevant thought forms incorrectly. The final step would be to amend their meta-models so that if these enhanced meta-models had been used from the outset, they would have produced models that did not require amendment.

When this second phase of model building is complete, the individuals would be required to repeat the process by testing their models against additional relevant sources of existing knowledge. This might include talking with experts in the field as a further test of the effectiveness of their models and meta-models. Eventually, it might include implementing strategies suggested by the models, and using the outcomes to assess the models further. And so on, recursively.

This process would be repeated with other complex problems that the individual is highly motivated to solve. In these new cases, the models that are built are likely to differ from each other. However, the meta-models that are used to scaffold the building of each of the specific models are likely to share greater similarities. Individuals would be required to generalise their meta-models as far as possible so that they could be used to build models that are effective for a range of complex challenges and problems.

This set of model-building practices that are directed at generating a capacity for self-scaffolding would also be useful for overcoming the second limitation in the design of the pilot program. This limitation arose from the need to adapt practices to the characteristics of each individual. The Pilot relied too much on group teaching. Nevertheless, it had been worth testing the effectiveness of group approaches. If it had proven effective, it would have assisted considerably with the spread of metasystemic cognition. One-on-one coaching considerably limits the number of new individuals a teacher can train in a given period.

However, individualised coaching is considerably more effective than group processes. It allows the coach to assess in detail the current cognitive level of the individual, assist the individual to see the specific limitations of their existing thinking, identify the particular movements in thought that the individual needs to incorporate in their thinking if they are to develop metasystemic cognition, and so on. The coach can use this information to design scaffolding that is adapted to the specific characteristics and needs of the individual. The nature of the scaffolding that proves optimal will differ considerably between individuals, and is likely to diverge further as the coach guides the individual in the recursive enhancement of their cognition.

When individuals achieve the ultimate goal of self-scaffolding, the particular way they go about it will likely also be unique for each individual.

The third main lesson that I learned from conducting the Pilot is that individuals need to be highly motivated. This is essential if they are to undertake the intense work on themselves needed to develop metasystemic cognition and the ability to be self-evolving. Up until this point in the evolution of humanity, very few individuals have developed the capacity to be awake in the midst of ordinary life. Fewer still have used this capacity to become self-evolving and to develop metasystemic cognition. Many, many more have set out on the path to awakening, but have died asleep. Few find the narrow road and the small gate that leads to awakening, let alone the road that is even less travelled that leads to metasystemic cognition.

As I have mentioned, Gurdjieff said that in order to awaken in the midst of ordinary life, an individual must work on themselves so hard that the soles of their feet sweat.

The development of these capacities will become easier as humans design and refine psycho-technologies that can train these capacities. These technologies will be enhanced considerably as an increasing proportion of people develop metasystemic cognition. This is because it is only with cognition at least at the level of metasystemic cognition that an individual can build mental models of the relevant cognitive and psychological processes, and use these models to develop complex psycho-technologies.

But it is early days yet. At present, in the absence of societal supports and institutions, individuals need to be highly motivated if they are to develop these capacities. The teaching strategy of getting students to work on their model-building skills by engaging with issues that matter to them most, will help to some extent. But alone, it will often not be enough. In general, if they are to succeed, individuals will have to have had life experiences and possess personal characteristics that already strongly motivate them to become self-scaffolders. In my experience, as yet there are very few of these individuals. The right combination of accidents of genetics, birth, upbringing, cultural influences, and so on, rarely occurs.

If an individual is highly motivated in this way, the information and feedback that they can receive in group sessions and/or by reading relevant books and other materials might be enough. They may have the mental energy and focus to use these sources of declarative knowledge to begin the process of self-scaffolding. Their efforts may become recursively self-reinforcing as they see their skills, models, and meta-models improving progressively.

Over the years since the Pilot, I have taken a few individuals through the modified version of the program in order to test its main features. In general, it has been significantly more effective than the group processes used in the Pilot.

One of these free-ranging guinea pigs, Mark Roddam, wrote a report on this experience that was published in the Integral Leadership Review.[16]

Given the personality characteristics I mentioned earlier and the existence of other priorities, I decided against devoting much more time and energy to promoting and implementing the new program. Instead, as I will outline, I returned to focusing on the further development of the evolutionary worldview and identifying in greater detail its implications for humanity.

In the meantime, I took steps to increase the likelihood that anyone interested in the scaffolding of higher capacities would encounter my contributions.

To this end, I made relevant materials freely available on my Evolutionary Manifesto website. These included details of The First Planning Meeting for the Second Enlightenment and an outline of the scaffolding program and its methods. I also talked about methods for developing higher cognition in several YouTube interviews. More recently, I participated in a series of online Salons about the scaffolding of metasystemic cognition organised by Bernhard Possert of the Center for Applied Dialectics.

I hoped that this would eventually attract contact from individuals interested in teaching the scaffolding program or their own version of it. I intended to supply additional teaching materials and assistance to any individuals who seemed committed to doing so.

However, no one has yet set out to implement the program or any other variant that they have developed themselves. Given that I am now over 70 years old, I thought that it was time to write down a detailed account of what I have learned from a lifetime of recursively improving myself and others. Hence this book.

---

[16] Roddam (2014) – see References for full citation

HUMAN SUPERINTELLIGENCE

18.

Origins of Life, Consciousness, and Global Governance

In the last ten years, I have had another half dozen papers published in international peer-reviewed science journals about various aspects of the evolutionary worldview. In general, the papers were all directed at strengthening the scientific foundations of the central pillar of the worldview: that evolution has a trajectory and that this has major implications for humanity and for how we organise ourselves collectively.

One of these papers set out to demonstrate that my approach (Evolutionary Management Theory) provides a better and more useful explanation of the major cooperative evolutionary transitions than its main competitor, David Sloan Wilson's group selectionism.[17] Two of the papers focused directly on identifying and justifying the existence of the evolutionary trajectory in greater depth and breadth.[18]

I decided to write two other papers because they had additional potential to increase acceptance within the scientific community of the foundations of the evolutionary worldview. Each of these papers used Management Theory to solve major problems that science had found intractable so far. If my theoretical framework was able to generate breakthroughs in these areas, the papers would raise the profile and reputation of my evolutionary work significantly. In turn, this would help generate wider acceptance of the evolutionary worldview.

The first of these papers used Management Theory to explain the origins of life.[19] I had first outlined this explanation in my 1995 paper,

---

[17] Stewart (2020) – see References
[18] Stewart (2014), and (2019b) – see References
[19] Stewart (2019a) – see References

237

Metaevolution. However, I had not bothered to devote a full paper to it because it was not central to the foundations of the evolutionary worldview and to my wider goals. This was even though I had always recognised that it was my most 'commercial idea' in the sense that if it were accepted by mainstream science, it would attract widespread attention. However, my overarching goal was not to achieve scientific recognition and fame. It was to contribute to the successful advancement of the evolutionary process.

In 2016 when I was reassessing my strategies in depth, I realized that if I could write a paper that solved the problem of how life originated, it would assist greatly in drawing attention to my other work. In particular, it would attract attention to Evolutionary Management Theory and its foundational role in the evolutionary worldview.

Similar thinking applied to the second of these two papers. It outlines a theory of the evolution and development of consciousness.[20] The paper explains the emergence of consciousness by identifying the architecture and functioning of a 'minimally complex' subsystem capable of giving rise to conscious experience. It then goes on to demonstrate that such a subsystem can provide substantial adaptive advantages to the organism that possesses it.

Such a subsystem enables an organism to achieve body-environment coordination in novel circumstances without substantial prior learning or innate capacities. The subsystem achieves this by assembling an image of the organism's position in relation to relevant features of the environment. Then, the subsystem inspects/examines/interprets this image to assess whether the movements it is producing are achieving the desired coordination.

The image ('the object') and the interpreter of the object ('the subject') comprise the minimally-complex, subject-object subsystem. The object is experienced by the subject. It 'lights up' for the subject. There is something it is like to be the subject.

In humans, threading a needle is a familiar example of conscious body-environment coordination. It is impossible to thread a needle in

---

[20] Stewart (2022) – see References

novel, unpractised circumstances without giving full conscious attention to your fingers, the thread, and the needle. Furthermore, it is impossible to imagine undertaking such an unpractised task if the relevant images are not 'lit up' in consciousness.

The paper goes on to consider how the further evolution of such a simple subject-object subsystem gave rise to the complex features of consciousness found in advanced organisms such as humans. In particular, it provides an account of the emergence of different levels of conscious modelling. It deals with the emergence and significance of metasystemic cognition and its spread in a Second Enlightenment.

Another paper I published dealt more directly with the implications for humanity of Management Theory and the evolutionary worldview, here and now. My goal was to draw on an understanding of evolutionary processes to identify how human societies need to be re-organised if humanity is to survive and thrive indefinitely. The paper outlines how human societies can be set up so that, thereafter, 'the good' will self-organise. Currently, human societies self-organise 'the bad'. Whether or not the members of a society intend it, our societies are currently self-organising environmental destruction, the threat of nuclear annihilation, poverty, and so on.[21]

The paper shows how an appropriate system of governance can be set up so that the interests of all members of society, including corporations, are aligned and remain aligned with the interests of the society as a whole. Furthermore, and of even greater importance, it shows how the system of governance can be set up so that it is highly evolvable, and the interests of the individuals and processes that constitute the governance system are also aligned and remain aligned with those of the society as a whole. This is essential to prevent the power of governance from being hi-jacked to exploit the society.

When these conditions are met, individuals, corporations, and others who pursue only their own immediate interests and have no concern for the interests of others will nonetheless act in the interests of society as a whole.

---

[21] Stewart (2018) – see References

HUMAN SUPERINTELLIGENCE

The paper argues that humans face an imperative to set up a global society that self-organises 'the good'. The survival of human civilization demands it. This will require the establishment of an appropriate system of global governance.

Of course, at the mere mention of global governance, there are many who will immediately reach for their gun. This is particularly the case if they are citizens of a current superpower, or are a citizen of an ally of one. It is generally in the interests of a superpower to oppose the erection of global governance that would have the power to manage and control all nation-states, including superpowers.

Superpowers do extremely well out of international competition, at least in the short term. They habitually use their power to out-compete other nations and to exploit less powerful ones. It is fundamentally against their interests to be subjected to a system of global governance that prevents destructive competition and exploitation and that promotes fairness in international relations.

Consequently, it is in the interests of those who have the most wealth and influence within a superpower to ensure that most citizens oppose global governance. Typically, they use their influence over the media and government to cause citizens to believe that global governance is their enemy and should be opposed at all costs.

It is also in the interests of the powerful to make citizens believe that their opposition to global governance has been freely arrived at and has arisen because such governance is objectively evil, not because their opposition has been orchestrated by media and government manipulation.

Often, the goal of the powerful is to ensure that citizens (but not themselves) are willing to die to prevent global governance.

Twice, humans on this planet have gotten close to instituting a system of global management that would disarm nation-states and prevent further war. After both the First and Second World Wars, major movements emerged to ensure that these kinds of abominations could never occur again. In both cases, there was general agreement amongst the majority of those charged with accomplishing this that global governance was essential for achieving and maintaining world peace. In both cases, the superpowers prevented this from happening. In both cases, sham global systems were set up that did not have the power to disarm nation-states and regulate their behaviour.

In the case of the First World War, the sham arrangement was the League of Nations, and in the Second World War, the sham was the United Nations. The citizens of superpowers now tend to believe that the failure of those institutions to prevent war between Nations is evidence of the ineffectiveness of global governance. They do not realize that this failure was instead due to the intentional actions of the superpowers at the time. The superpowers set up the League of Nations and the United Nations so that they would fail. Most citizens of the world were duped, and then duped again.

The consequences of these intentional actions by superpowers came extremely close to facilitating the destruction of human civilization. When the United States was blockading Cuba in order to prevent the deployment of Russian nuclear missiles in Cuba, the world was saved from nuclear conflagration by the actions of one officer on a nuclear-armed Russian submarine. The submarine was being depth-charged by the blockade, and the crew thought that they were doomed. The rule that regulated the firing of nuclear weapons by the submarine required that the three most senior officers had to agree unanimously before this could be done. Two of the officers believed that World War Three had started, and that they should nuke United States cities. One disagreed, and vetoed the use of their nuclear weapons.

If this Russian had voted in favour, it is likely that no one would now be around to ensure that future generations would know of the criminal culpability of the superpowers who had stood in the way of effective global governance. Nor would they know of the criminality of the wealthy and powerful who had influenced the actions of superpowers on these issues. They would not know of the culpability of the functionaries who were used by the powerful and who allowed themselves to be so used.

With an effective system of global governance, disputes like the Cuban missile crisis would be settled by law, not by war. They would be resolved in the same way that disputes between states within the United States of America are decided: by law, not by war.

It is extraordinary that the evolution of human civilization on this planet came so close to being terminated. But the threat of nuclear annihilation has not diminished since then. At present, there are far many more nuclear weapons on the planet than then, and many more nation-states have them.

Various false beliefs that purport to justify opposition to global governance are commonplace among the citizens of superpowers. In particular, many believe that a global system of governance, which they often personify as a Global Government, will exploit the people of the world and remove their freedoms.

For example, few of the citizens of the United States see that the creation of a system of global governance would merely involve the repetition at the level of nation-states of a process that happened many years ago to produce the United States of America. Many of the States of North America were brought together by a system of federal governance to produce the United States. This overarching governance was essential for enabling the emergence of what many Americans believe is the greatest country on Earth.

Few attribute the enormous benefits produced by this unification to the establishment of a higher level of governance. They do not see the direct benefits produced by this governance: It enabled the disarming of the States and rendered unthinkable the war and destructive competition between the States that was prevalent before the creation of the United States.

Few see that the same is just as feasible and just as beneficial at the international level.

Unfortunately, even fewer have the cognitive capacity to envisage how a global system of governance could be organised in such a way that the interests of the system of governance are aligned with the interests of the citizens of the global society. Not only can they be aligned, but as is the case with all living organisms, the society can be organised in such a way that any disruption to these arrangements will be repaired automatically. The global entity will self-organise any repairs to the system of global governance that are needed to ensure that the alignment of interests is maintained.

When this form of global organisation has been implemented, the individuals and processes that constitute the system of global governance will, when pursuing their own immediate interests, advance the interests of the society as a whole.

Properly constrained, global management will massively increase the opportunities and freedom available to humans across the planet. It will achieve this by ending war and exploitation by the powerful, and by significantly increasing the opportunities available to citizens. These opportunities will be produced by the explosion of internal differentiation which will accompany the emergence of the complex cooperation that is enabled by global management.

These kinds of processes and arrangements can guarantee that the overwhelming benefits of global management can be achieved without falling prey to dangerous exploitation and restrictions of freedom. Great power has no intrinsic danger, provided it is appropriately constrained. It is only unconstrained power that is a problem.

Past evolution provides us with many examples of forms of organisation that constrain the power of management in order to align its interests with those of the managed organisation as a whole. Selection driven by competition between managed organisations will favour the emergence of systems of constraint that prevent managers from exploiting their organisations. This is because well-managed organisations can outcompete organisations that are hampered by managers that exploit the organisation, reducing its productivity.

Examples that are easiest to envisage are those at the level of human societies. The signing of the Magna Carta in Britain in 1215 was an event not just of historical but also of evolutionary significance: It implemented a series of constraints on the king's power, aligning his interests more closely with those of the nobles.

The emergence of democracy saw a further significant increase in the constraints that applied to governments. It was a major innovation in mechanisms that align the interests of governments with those of the governed.

Another important innovation was combining systems of constitutional governance with the separation of powers. The constitution is a set of rules that constrain the governors, and it is enforced by an independent judiciary, not by the governors.

At the global level, similar arrangements can be adapted and evolved in many different ways in order to embed global governance in an appropriate system of constraints.

My paper about how to set up a society that self-organises 'the good' also takes a novel approach to dealing with these issues. In particular, it identifies how new kinds of mechanisms can be established that drive the self-organisation of governance that is in the interests of the governed.

The paper suggests that this can be achieved if a society is organised so that all the members of the society (including individuals, corporations, governments, etc.) capture the impacts of their actions on the society as a whole. For example, if their actions harm society, they will experience harm. If their actions benefit society, they will capture the benefits. I refer to this condition as 'consequence capture'. Once it is implemented, the interests of all members of a society will be aligned with those of the society as a whole, and the pursuit by individuals of their immediate interests will benefit the society.

How might 'consequence-capture' be implemented?

The paper begins by considering the basic management architecture that is common to all forms of governance. It then goes on to identify what additional arrangements would be necessary to ensure that improvements in governance self-organise inexorably. The paper argues that this can be achieved if governance is established by a distributed, invisible-hand system of exchange relations that functions in a similar way to the 'invisible hand' that currently self-organises our economic markets.

I refer to such a distributed market that deals with improvements in governance as a vertical market. Taken together, the vertical market in governance combined with a horizontal market in goods and services will produce a self-organising society.

Broadly, the vertical market is a market in which 'producers' of possible improvements in governance can 'sell' particular improvements to citizens who can 'buy' any that they want. Typically, proposed improvements in governance will impact groups of citizens (governance exists inherently to provide collective benefits. In contrast, economic markets typically deal with goods and services whose effects can be restricted to a particular purchaser.) Consequently, the potential buyer in any instance will be the citizens who will be impacted by the proposed improvement in governance.

The self-organising nature of such a market is easy to see: any current failure in governance will be a profit opportunity for 'providers' and 'manufacturers' of governance. The vertical market system will incentivize them to search for profit opportunities and to invest in research that develops improvements in governance that can correct the failures. Producers of governance will compete to design and offer elements of governance that overcome deficiencies.

Viewed through the lens of this distributed, self-organising framework, our current system of democracy is an absurd and incompetent method of establishing and adapting governance. Imagine how unresponsive and ineffective economic markets would be if they operated under similar principles—i.e., if consumers had to purchase all their goods and services in advance for a four-year period through a collective decision that can choose one of only two or three packages of goods and services.

If economic markets were as restricted and limited in these ways as political markets, they would not have given rise to the enormous diversity of goods and services tailored to individual needs and wants that we see in modern societies. They would be just as incapable of aligning the interests of producers and consumers as are current democratic systems at aligning the interests of the governors with those of the governed.

It is not surprising that current systems of government are as inept and thoroughly incompetent as I discovered first-hand during my working life. But this is just a temporary evolutionary phase that we are passing through. With the spread of metasystemic cognition will come the ability to organise much more effective and creative ways of establishing and evolving systems of governance.

19.

The Power of the Evolutionary Worldview

I indicated earlier in the book that when I get toward the end of outlining the history of my cognitive development, I will provide a brief overview of the evolutionary worldview and its implications for humanity. I have reached that point.

This overview will provide an outline of the evolutionary processes that have shaped life on Earth in the past 3.5 billion years and that will continue to do so in the future. As I have mentioned earlier, an understanding of these evolutionary processes identifies major implications for how we must organise ourselves and adapt ourselves psychologically if humanity is to survive and flourish indefinitely into the future. Previously, I have also emphasised that these implications point to what humanity must do here and now, not in the far-flung future.

This sketch of the implications for humanity of an understanding of the large-scale processes in which we are embedded is highly relevant to the book's goals. It serves as a demonstration of the power and necessity of metasystemic cognition. The analytical/rational thinking that underpins current mainstream science cannot build adequate mental models of the complex challenges that humanity faces. It is ineffective at predicting how the complex processes that will shape our future will evolve and develop.

Equipped only with analytical/rational thinking, we fly blind into a complex and dangerous future. Without metasystemic cognition, we cannot see enough of our future to identify how we need to act, here and now, in order to avoid being selected out of existence.

Without metasystemic cognition, we cannot understand the trajectory of evolution and the related processes that shape our future.

Metasystemic cognition is essential for building a dynamical map that can guide us. Without metasystemic cognition, we have no evolutionary worldview and no map. We will be helpless before powerful forces that we cannot understand and are beyond our control.

For these reasons, this sketch of key features and implications of the evolutionary worldview should be seen as a demonstration of the necessity for metasystemic cognition. Together with my more detailed publications referenced in the footnotes, this outline also serves as a practical demonstration of the application of metasystemic cognition to a set of complex phenomena.

As such, it can be used as a bundle of resources for generating exercises that will help you to recursively improve your own cognition towards the metasystemic level. I have already mentioned some of the essential elements of such a process of recursive self-scaffolding. But in the book's final chapters, I will set out in a more straightforward and comprehensive fashion how you can engage in such a process.

This outline of the evolutionary worldview will draw together various strands that have already been considered to some degree earlier in the book. To this extent, it will contain some repetition. But as you will see, it is essential and powerful to combine all these strands so that they can be viewed as a coherent evolutionary whole.

The trajectory of evolution has exhibited three over-arching trends as it unfolded during the past three and a half billion years on Earth. First, living processes have diversified as evolution proceeded. They have also been integrated progressively into cooperative organisations of ever-increasing scale. As this has occurred, living processes have also tended to increase in evolvability. Growing diversification, progressive integration, and increasing evolvability are the three over-arching trends that have characterised the trajectory of evolution on Earth.

When life first emerged from chemistry, each living entity was an infinitesimal cooperative organisation of molecular processes. From its very beginning, life has been fundamentally cooperative and integrative. These cooperatives of molecular processes gave rise to the first simple cells. Through the operation of standard evolutionary processes, these simple cells soon diversified into numerous forms that were each adapted to particular environments. Life quickly spread across the planet, specialising and differentiating as it went.

The first major step in the integration of living processes occurred when some simple cells banded together to form cooperatives. These cooperatives gave rise to the complex eukaryote cell. Our bodies are composed of eukaryote cells. The mitochondria within these cells are the recognizable descendants of bacteria that participated in this initial integration process.

Propelled by the advantages of cooperation, eukaryote cells diversified across multitudes of environments.

The next major integrative step was the emergence of cooperatives of eukaryote cells. These diversified across the planet to produce humans and the other multicellular organisms that surround us today.

As is the case with all these cooperative transitions, the cells that were integrated into multicellular cooperatives were released from the need to perform all the functions necessary for survival and flourishing. This freed them to specialize in ways that served the cooperative's needs. Consequently, we find significant diversification and specialization among the cells within multicellular organisms, including within our bodies.

In a repetition of the pattern underlying these first two major transitions, cooperatives of multicellular organisms emerged, spread, and diversified. These manifested as insect societies and eventually as cooperative groups of some mammals, including primates.

This pattern was again repeated with the evolution of cooperative groups of humans. The first of these were kin groups, e.g., a family of individuals who were very closely related genetically. Bands and then tribes emerged once evolution had solved the problem of how to form stable cooperatives of unrelated individuals who previously competed destructively with each other. These were followed by cooperatives of tribes that eventually produced agricultural communities, kingdoms, and city-states. Cooperatives of these produced empires and, eventually, the modern nation-state.

This process of integration began with cooperatives of molecular processes that were less than one-millionth of a meter in diameter. In a step-wise process of increasing integration, it has now produced Nations and cooperatives of Nations that span continents.

At each step in the integrative process, groups of smaller-scale entities formed cooperatives that became larger-scale entities at the next level.

This progressive integration is driven by the adaptive benefits of cooperation. At every level of organisation, cooperative teams have the potential to out-compete isolated individuals that live and act alone. Cooperative groups have the potential to develop greater power, command more resources, act over greater scales, and become more evolvable. They can also take advantage of the efficiencies and synergies that are enabled by a division of labour and specialisation.

Before integration emerges at a particular level, the smaller-scale entities compete against each other. Until appropriate management emerges, smaller-scale entities that are cooperators will tend to be out-competed by entities that do not cooperate and that free-ride on the cooperation of others. This prevents the emergence of complex cooperation, no matter how potentially beneficial the cooperation might be.

As we have seen, this barrier to the emergence of cooperative organisation can be overcome by an effective form of powerful management. Such a manager has the power to reach across the organisation to punish free-riders and support cooperation. Effective management will align the interests of the smaller-scale entities with the manager's interests.

Competition between managed organisations will then tend to align the manager's interests with those of the organisation as a whole. As a result, smaller-scale entities that pursue only their own immediate interests will thereafter act in the interests of the large-scale organisation. Within emerging larger-scale organisations, cooperation will pay. Free-riding will not.

Once managed organisations emerge, continued competition between them will drive an entification process. Given that the interests of both the manager and the smaller-scale entities are all aligned with those of the emerging entity as a whole, selection will favour adaptations that equip large-scale entities with the capacity to adapt and evolve as coordinated wholes.

Increasingly, this will enhance the ability of the large-scale entities to act, adapt, and evolve as coherent entities in their own right. For example, in the cooperative groups of cells that eventually produced humans, this process of entification drove the emergence of brains, nervous systems, and adaptive capabilities.

If there had been no barrier to the emergence of beneficial cooperation, evolution on Earth would not have taken over three billion years to get to the current point where integration is being achieved on the scale of continents. Instead, it would have proceeded much more quickly until integration had occurred on the scale of the planet as a whole. It is only the evolutionary impediments to the emergence of complex cooperation in unorganised groups that has prevented the rapid emergence of integration.

Currently, evolution on Earth has reached the point where integration is just one step away from occurring on the scale of the planet, potentially producing a unified, cooperative, sustainable, and highly-evolvable global entity. At present, however, we are still at the stage where smaller-scale entities in the form of nation-states and multinational corporations continue to compete destructively. They have not yet integrated into a large, cooperative whole.

This destructive competition manifests in the form of environmental degradation and the threat of nuclear war. The entities involved in this competition are the largest in scale, power, and evolvability that have ever existed so far during the evolution of life on Earth. As such, their destructive competition has the potential to threaten the continued existence of human civilization and many other living processes on Earth.

As I have discussed earlier, a proper understanding of the past evolution of life on Earth identifies how these existential threats can be overcome. The same organisational architectures that have terminated destructive competition and promoted cooperation at lower levels, can also do so at the level of nation-states. Appropriate management can ensure that destructive competition is no longer in the interests of nation-states, multinational corporations, or other entities below the global level.

This management can establish a highly-evolvable system of governance that ensures that any actions that act against the interests of the global organisation are disincentivized. No longer would it be in the interests of any entities to act in ways that are environmentally destructive or that could otherwise contribute to existential threats.

Of course, as I have discussed previously, it would be essential to ensure that the individuals and processes that constitute the system of global governance are also appropriately constrained. These constraints are necessary to ensure that their interests are aligned with those of the global society as a whole. This will prevent exploitation and abuse of power by global management.

The great challenge that faces the evolutionary process when it reaches this stage is that natural selection will no longer exist to drive the emergence and entification of the global entity. Natural selection is driven by competition between the members of a population of entities. However, once the integration process has progressed to the global stage, there is only one entity at the planetary level. There is no population of competing entities that will drive the selection of global entities that are better adapted and more evolvable.

In the absence of selection generated by a population of global entities, entification is likely to occur only if it is driven intentionally by humans. In order to identify what needs to be done to produce entification, humanity will need to draw upon its understanding of previous entification processes at lower levels, and also its knowledge of the trajectory of evolution. This in turn will require cognitive capacity at the metasystemic level.

Why might humanity be motivated to do this? Individuals with the requisite cognitive capacity will see that doing so is essential for the survival and flourishing of life on Earth, now and into the future. It is only as a fully-developed global entity that humanity will be able to participate in and contribute positively to the future successful evolution of life in the universe. To stop short of producing a fully-developed global entity would be like a baby stubbornly refusing to leave the womb. The consequences would be just as disastrous.

If the evolution of life on Earth is, as it seems, a developmental process that is set up to produce a global entity, why would it be set up so that its successful completion depends on humans developing higher-level cognition and deciding to complete the process intentionally? If this developmental process has been shaped by larger-scale selection, why would all this not just be pre-programmed?

However, this is exactly what might be expected of a developmental process that has been shaped and tuned by larger-scale evolutionary processes that act at the extra-planetary level. None of the developmental processes that produce living organisms are set up to proceed mechanistically, with each and every step in the process pre-programmed. If they were, developmental processes would be extremely fragile—any disturbance would likely disrupt the process seriously. For this reason, all developmental processes make use of the adaptability of the entities that constitute the developmental process. The actions that the entities need to take in order to produce development are not pre-programmed in detail.

A development process that comprises constituent entities that possess metasystemic cognition will have a much greater potential to develop a complex and evolvable larger-scale entity than will entities at the analytical/rational level. For this reason, successful developmental processes will tend to make use of the highest capacities of which their constituent entities are capable. It is, therefore, no coincidence that the developmental process in which we are embedded is demanding that humanity transition to metasystemic cognition.

Broadly, entification at the global level would require the establishment of a highly-evolvable system of global governance/management. This global governance would need to manage not only humanity and organisations of humans, but also all technologies (including AI), all other living processes on the planet (including ecosystems), and relevant bio-geo-chemical cycles and systems. In general, all the living and non-living processes of the planet that are relevant to the survival and flourishing of the global entity would need to be managed to serve the interests of the entity.

The processes that constitute global governance would also comprise AI and other relevant technologies, as well as humans and the other living and non-living processes that support them.

Furthermore, global entification would require the development of structures and processes that enable the global entity to act, adapt and evolve as a coherent whole. Using the terminology developed by the great systems thinker Stafford Beer, initially this would mean developing the capacity to adapt and evolve 'for the inside/now'. In effect, this would involve developing internal processes that regulate and organise the 'metabolism' of the global entity. These include the economic systems and market processes that function within the entity. In turn, these include processes that acquire and distribute energy and other raw materials, and also the processes that determine which particular goods and services are produced internally.

At present, some markets that function on a global scale have emerged already. However, the emerging global system currently has little capacity to adapt for what Beer refers to as 'the outside/future'. This is the capacity of an entity to adapt in relation to events that occur external to the entity, including events that occur in the future. In general, this entails the capacity of an entity to achieve its goals by acting upon and modifying its external environment. A simple example of a global entity adapting for the outside/future would be a case in which the entity detects an asteroid that is about to collide with Earth, and then destroys the asteroid before it does so.

For an emerging global entity to adapt and evolve for its outside/future, it needs the ability to act and adapt as a coherent whole. In order to adapt in relation to future events, an entity also needs the ability to form models of the future consequences of its actions, and to use these models to identify the actions that will adapt it optimally, given future predictions.

Of course, the individual humans that constitute the global entity will not need to know how they should adapt individually in order to produce the optimal adaptation of the global entity. Instead, they will be like cells in our bodies. Our cells do not know us, care about us, or understand our needs. They simply pursue their cellular needs and goals. Yet, their collective actions produce our speech, movements, and so on.

This occurs because once our brains decide how we should act in a particular situation, they initiate actions that change the environments of the relevant cells. These changes ensure that when the cells adapt to them, they collectively produce the desired adaptation of our bodies as a whole.

In a similar fashion, the management of the global entity will embed citizens and organisations of citizens in a system of incentives and disincentives. These will evoke the actions needed to adapt the global entity. By pursuing their immediate human interests, citizens will serve the adaptive interests of the global entity.

From the perspective of individuals, citizenship of a global entity will increase the range of choices available to them, not restrict them. It will not reduce their freedom. They will continue to be able to decide freely what career to follow, what work to do, and how to spend their income. But now they will be able to choose freely from a much greater array of options. The internal diversification that will emerge within a global entity as it develops will massively multiply the available choices.

A key difference from the present is that whatever work they choose, citizens of the global society will know that their job will serve the interests of the global society as a whole. This obviously cannot be said of many of the work opportunities offered in our current societies. Currently, many jobs involve work that is helping to drive existential crises that threaten human survival and flourishing.

Nor will the global entity impose uniformity on its citizens. An organisation's adaptability and creativity depend on the diversity of its members' perspectives, skills, and capacities. Hence, the global entity will enhance its effective diversity by ensuring that the potential of all citizens is fully realized, no matter their ethnic and cultural backgrounds. Universal education, minimum income, and health care across the entire planet will be some of the key foundations of any global entity that is set up to maximize its evolvability.

Without natural selection generated by competition between global entities, the development of a capacity to adapt for the outside/future will not be driven automatically. It will emerge only as a result of the conscious action of its members. If a global entity is to develop this capacity, the complex global structures, systems, and processes that enable it will have to be built intentionally by humanity, at least initially. Given the dynamical complexity of the tasks needed to achieve this, it will also require cognition at least at the metasystemic level.

Once these enabling arrangements are in place, the global entity will be able to adapt as a coherent whole. It will be able to implement particular actions by orchestrating appropriately the behaviour of its constituent entities.

Eventually, in order to enhance its evolvability, the global entity will need to develop the capacity to recursively improve its own cognitive capacities. It will also have to become a self-evolving being. This will necessitate it freeing itself from the dictates of its evolutionary past, including from the evolutionary pasts of the humans that constitute it. This is likely to require the use of meditation-like processes that enable it to dis-embed from lower-level processes in order to build a higher-level self that is free from the constraints of lower levels. The global entity will recursively develop the capacity to re-make itself repeatedly to meet whatever evolutionary demands it foresees in its future. An important component of its capacity to envisage likely future events will be its theories of evolution, particularly its mental models of the future trajectory of evolution.

Can the existence of the trajectory of evolution provide meaning and purpose for human existence?

The processes that produced the trajectory of evolution on Earth are not likely to be unique to this planet. It has become increasingly evident that there are many millions of planets elsewhere in the universe that appear to be as life-friendly as Earth. Furthermore, an understanding of what it actually takes for life to emerge suggests that it is not likely to be extremely rare and improbable. All that is needed is the emergence of collectively autocatalytic chemical processes that become to be managed by, for example, RNA-like molecules.

The identity of the particular molecules that instantiate this management architecture may differ significantly among planets on which life emerges. But the architecture is likely to be the same.

This can also be said about subsequent major cooperative evolutionary transitions. The processes that have driven these transitions on Earth are not unique to Earth. They do not require anything rare or improbable. Like the emergence of life itself, all they require is the emergence of groups of entities that are somewhat collectively self-producing, and the emergence of powerful managers that have the capacity to manage groups in ways that comprehensively overcome the barrier to cooperation.

Again, the architecture of each subsequent cooperative transition is likely to be similar across planets on which life emerges, but the details of the entities involved in the transitions are likely to differ.

Importantly, the trajectory of evolution on each planet seems likely to culminate in the emergence of an entity that encompasses the planet. Again, the processes that are proceeding towards this outcome on Earth do not appear to be peculiar to Earth, nor do they seem rare or improbable.

What is going on in our universe?

Imagine that we could observe this process unfolding across many planets in our galaxy. At different times, we would see life emerge on particular planets. Or single cells might arrive from elsewhere. Wherever life commenced on a planet, we would see it diversify across the planet. Then, we would see it progressively become integrated in a step-wise process. Eventually, we would witness the emergence of a living entity on the scale of the planet. This global entity would undergo a process of entification, developing the capacity to adapt and evolve as a coherent whole.

Now imagine observing these events unfold across planets, but with the process sped up so that it takes only an hour or so of our time for the full evolutionary trajectory to complete itself on any given planet. In the overwhelming majority of instances in which we see life begin on a planet, we will then see it develop rapidly to produce a living entity on the scale of the planet. As we watch, the evolutionary process on many planets will soon hatch a global entity.

Overall, we will witness a process that looks like it has been set up to develop and hatch global entities on suitable planets. The process that unfolds on a particular planet will look like an embryo that commences as a single cell and then progressively develops structures and processes of larger and larger scale. These arrangements are destined to eventually perform particular functions within the global organism that develops from the embryo. It will appear as if what we are observing is a developmental process that has been pre-programmed to eventually hatch an organism on the scale of the planet.

Equipped with metasystemic cognition, we can zoom out in time and space, and observe how these processes will unfold further into the future. We see that once global entities develop the capacity to adapt and evolve for the outside/future, they begin to communicate with each other (fully-developed global entities will not attempt to communicate or interact with life on a planet that has not yet produced a global organism. In order to develop the structures and processes that are needed if a global entity is to develop fully, the components of a developing entity will need to be conditioned by the developmental challenges that they will encounter at earlier stages. Contact with advanced entities could circumvent and prevent this essential learning. For example, it is highly unlikely that humanity would go through the long and difficult process of developing an effective form of global governance unless destructive competition at the level of nation-states demands that we do so. This is not a process that can be easily circumvented through external intervention.)

As we continue to watch, we see cooperative organisations of global entities emerge and evolve. These organisations of planetary entities themselves undergo entification processes and form yet larger cooperative organisations. And so on. The goals of these larger-scale entities are as unimaginable to us as our love lives to the bacteria in our gut.

We see that when dealing with other global entities, entities are guided by the evolutionary worldview that they used initially to guide their emergence and entification. As we have seen, this worldview indicates that evolutionary success depends on entering into cooperative relationships of ever-increasing scale. Furthermore, it demonstrates that doing so can be achieved safely and without self-sacrifice by establishing higher-level management that facilitates cooperation and suppresses free-riding and defection.

Eventually, we see that the universe becomes infused with life and hatches an agentic entity on the scale of the universe. It is possible that entities at this level might reproduce universes. The resultant natural selection between universes might constitute the larger-scale selective processes that, as mentioned earlier, might shape and tune the developmental processes within universes that hatch global entities and ultimately produce entities on the scale of the universe.

However, at this stage, these are only possibilities. Evidence does not yet exist to test these or other possibilities.[22]

Equipped with metasystemic cognition, we can also zoom in to examine more closely particular phases, events, and challenges that are common across planets as life develops on them.

We begin by focusing on the emergence of organisms such as humans that develop the capacity to form mental models of their environment and their interactions with it. We see that on each planet on which such organisms emerge, they typically undergo something like the First Enlightenment in which many of them develop the capacity to form mental models that include abstract representations as well as ones that are concrete (in Piagetian terms, they transition from the concrete operations level to the formal operations level. In the terminology I am using, the formal operations level equates broadly to analytical/rational thinking). As we watch, we see this transition give rise rapidly to science and technological development, including industrialization.

Eventually, these intelligent organisms have their Darwin: one or more individuals use their capacity for abstract/rational modelling to

---

[22] Stewart (2010) – see References

develop a basic theory of biological evolution driven by natural selection.

As we continue to watch, this new evolutionary perspective combines with analytical/rational thinking to begin to undermine religious belief systems. Previously, these belief systems orchestrated the organism's behaviours, including in ways that tended to produce viable social systems. We see that the undermining of religious beliefs tends also to undermine the effective functioning of their social systems.

Eventually, we begin to see the emergence of individuals with metasystemic cognition. As metasystemic cognition spreads, increasing numbers of individuals are able to construct mental models that enable them to understand the kinds of complex dynamical architectures that need to be put in place to ensure that their societies function effectively for all their members. The organism now becomes capable of building effective social systems consciously and intentionally, without the aid of religious beliefs.

Increasingly, we see individuals emerge who have developed and embraced a complex evolutionary worldview. For them, this changes everything. They no longer see themselves as isolated individuals who live for a short period in a universe entirely indifferent to their existence. Instead, they see themselves and their societies as having been produced and shaped by larger-scale evolutionary processes that will continue to shape life on their planet in the future.

As we continue to watch, they begin to realize that the evolutionary process that is unfolding on their planet looks very much like the processes that unfold as an embryo develops. It is as if life on their planet has been set up to eventually give birth to a living entity on the scale of the planet. They see that this entity will go on to develop its own agency, acting and adapting as a coherent whole. We also see the realization spread that these evolutionary processes are likely to unfold in a similar way on planets elsewhere, wherever life emerges.

But as we watch, an even more significant and powerful realization emerges and spreads amongst the organisms that are equipped with metasystemic cognition. This realization changes how these organisms understand themselves, and also what they do with their lives.

Their astonishing realization is that the developmental process will succeed in hatching a global entity only if they choose to do what is necessary to produce this outcome.

They see that up to this point, the evolution of life on a planet unfolds automatically, driven by blind natural selection. But if it is to progress beyond this, it must be driven intentionally.

We continue to watch as their growing understanding of evolution causes them to realize that they have specific roles to perform in the developmental process. These roles are critically important. If they do not intentionally fulfill them, life on their planet will be like an egg that goes rotten and never hatches. Life on their planet will have no relevant role in the future evolution of life in the universe. It will not hatch a global entity, it will not participate in communities of global entities, and so on.

We watch as a significant proportion of the individuals equipped with metasystemic cognition make the transition to intentional evolution. They commit to being evolutionary activists who will do whatever they can to advance the evolutionary process on their planet and ensure that the developmental process is completed successfully.

We see these individuals realize that they need to develop a proto-global entity as a matter of urgency. At this stage in the evolution of life on their planet, fewer than 300 large-scale societies exist. They know from their understanding of the trajectory of evolution that destructive competition inevitably emerges between the largest-scale entities that exist at any time. At this stage in the evolution of life on their planet, this destructive competition is powered by advanced technological development. As we watch, destructive competition drives environmental degradation, the threat of global war, and other existential threats. The organisms see this also.

We see the organisms adopt as a priority the goal of implementing a system of global management that will terminate this destructive competition and promote global cooperation. They act as if there is a race between the organisation of a proto-global entity and the destruction of their civilization.

As part of their strategy for actualizing a global entity, they realize that they need to spread their evolutionary worldview as quickly as possible. They realize that metasystemic cognition is a prerequisite for understanding a complex evolutionary worldview and its implications for what they must do.

As we watch, they also realize that if they are to actively implement strategies designed to advance the evolutionary process, they will need to become self-evolving beings—they will need to free themselves from the constraints of their genetic, cultural, and social past and to find satisfaction and motivation in whatever actions they need to take to advance the evolutionary process.

We see that initially, strategies for moving towards a system of global governance attract strong resistance. In particular, this opposition comes from those who benefit most from the continued dominance of the most powerful societies. These wealthy and powerful individuals have used their influence and dominance to have their societies impose a 'rules-based order' at the inter-societal level that enables them to continue to exploit other societies.

However, as we watch, increasing numbers of individuals develop metasystemic cognition and embrace an evolutionary worldview. Those with metasystemic cognition quickly achieve disproportionate power and influence. Their far superior ability to strategize effectively in complex circumstances enables them to out-maneuver those who are limited to analytical/rational thinking. This is the case no matter how intelligent the analytical/rational thinkers are at their level. Those with metasystemic cognition see and consider things that are relevant but that others are blind to.

We watch as relatively small numbers of individuals with metasystemic cognition are able to manipulate societies to ensure that they implement pro-evolutionary strategies.

Soon, we see the institution of a cooperative global society underpinned by global governance. This ends the perilously dangerous period in which destructive competition between technologically

262

powerful societies drove existential threats. We watch as those guided by an evolutionary worldview begin to work on building the evolvability and agency of the emerging global entity. Those with metasystemic cognition become cells in the brain of the global entity.

We zoom out again and shift our focus to other planets. We focus on those that have reached the perilous period in which destructive competition between sub-global societies is driving existential threats.

In the overwhelming majority of instances that we observe, metasystemic cognition and the evolutionary worldview spread sufficiently to institute a global system before destructive competition produces irreversible damage. But in a small number, the race is lost. We watch as the developmental process on these planets fails. They do not hatch a viable global entity.

The impact of destructive competition on these planets destroys the complex technological civilizations that have emerged. We see that the accumulated fossil fuels that 'pumped up' this complexity have been used up, and these civilizations are no longer sustainable. Furthermore, we see that there is no rebound effect. Life has used up the readily accessible energy sources essential to produce complex technological civilizations. The yolk that was needed to fuel the development of the embryo has been used up.

*   *   *

What has all this got to do with you?

At this time, on this planet, evolution has reached the perilous stage in which powerful sub-global entities compete destructively. These competing entities have the scale, power, and technological means to seriously damage the ability of Earth to sustain complex life. Unchecked, this destructive competition between nation-states, multinational corporations, and other powerful entities seems likely to end human civilization this century.

The rise of analytical/rational cognition has killed God and eroded the pro-social benefits of religious systems. Somewhat paradoxically, many of the most dangerous people on the planet claim to be religious or pander to others who hold religious beliefs.

263

HUMAN SUPERINTELLIGENCE

Almost no individuals or organisations have yet developed metasystemic cognition. Consequently, nearly all people and organisations are unable to develop effective mental models of the competitive dynamics, economic processes, political systems, and other complex processes that are driving existential threats. Consequently, almost no one on the planet at this time is equipped to identify the kinds of complex strategies needed to end the destructive competition that threatens the survival of human civilization.

Due to the extreme rarity of metasystemic cognition, almost no one on the planet has embraced a complex evolutionary worldview that provides an understanding of the large-scale evolutionary processes that have shaped life on Earth and that will determine its future. Very few see that the only viable way out of the perilous period in which we find ourselves necessitates the establishment of a unified, cooperative, and highly-evolvable global entity that is underpinned by appropriate global governance.

Even fewer individuals have done the considerable work on themselves necessary to become self-evolving beings. Those who have not done so find that their efforts to advance the evolutionary process on Earth are continually undermined by conflicting genetic, cultural, and social predispositions implanted in them by past evolution and conditioning.

As was the case for intelligent organisms on the other planets that we have imagined, all humans alive during this dangerous period who become aware of the evolutionary worldview are faced with difficult choices. These are choices that you face.

You now know that you and your fellow humans are embedded in larger processes that are headed somewhere and that are not random and meaningless. You are embedded in a developmental process that is directed at producing a highly-evolvable global entity. However, you also know that this process will not be completed successfully and will not hatch a viable global entity unless sufficient humans wake up to an evolutionary worldview. They must then commit intentionally to devoting their lives to doing whatever is necessary to help develop a living entity on the scale of the planet.

This can include contributing to the spread of the evolutionary worldview and producing an evolutionary awakening (the Second Enlightenment that I have referred to will be an Evolutionary Enlightenment if these efforts are successful); working on yourself to fully develop metasystemic cognition and the capacity to be self-evolving; contributing to the spread of these higher capacities to others; contributing to the development and implementation of strategies that will institute an appropriate system of global governance that will organise a proto-global entity; and working on establishing systems and processes that will enhance the evolvability of the global entity and that will eventually enable it to participate in communities of global entities.

Will you commit to a life dedicated to this evolutionary work? Will you make the transition to becoming an intentional evolutionary, an evolutionary activist? Will you work on yourself in order to make yourself a more effective intentional evolutionary? Will you, for example, embark on an arduous process of recursive self-improvement to develop metasystemic cognition and to become self-evolving? And so on? And so on?

Or will you continue to act in ways that contribute to the continuation of 'business as usual', becoming an accessory to the likely end of human civilization this century?

Do you find the required transformations of yourself as an individual and of the society in which you are embedded, too daunting and overwhelming?

Exposure to the evolutionary worldview will tend to erode your motivation to spend your life like a 'normal' human being. Currently, as I have mentioned, humans endlessly pursue the positive feelings produced by experiences such as popularity, self-esteem, sex, feelings of uniqueness, power, eating, and social status. They strive to avoid the negative feelings that accompany experiences such as stress, guilt, depression, loneliness, hunger, and shame.

Informed by an evolutionary understanding, you will find it increasingly difficult to experience meaning or purpose in a life lived in this way. You will see that actions motivated by these desires and emotions are absurd. These predispositions were shaped by past

265

evolution to cause you to act in ways that were adapted to the circumstances that existed then.

To continue to live your life pursuing desires that you now know are often maladaptive and that may contribute to the demise of human civilization will make your life an absurdity. You will be like a toy soldier that marches across the floor until it walks into a wall and falls onto its back. It continues to march on and on, going nowhere, until its battery dies. It does not have the capacity to adapt to its changed circumstances.

The main difference between you and the toy soldier is that it does not have a choice. It cannot construct mental models of alternative actions and choose the most adaptive ones.

If you have had the misfortune to read and understand this book, you will never be able to 'un-see' the absurdity of your life up to this point. If you continue to live such a life, you will be disturbed until you die. The easier option will be to undertake the relevant evolutionary transitions.

20.

Further Work on Myself

I was extremely lucky in 2017 and again in 2023 to encounter circumstances that strongly incentivized me to work on myself more seriously than I had ever done before. In both instances, it became a matter of life and death.

I mentioned earlier that life was generally very easy for me. I could use my abilities to get whatever I wanted. I was a master at strategically avoiding circumstances that were likely to impact me negatively.

This changed fundamentally in 2017. My youngest daughter Libby left a difficult relationship, taking her 9-month-old son William with her. But under Australia's family law system, this did not mean that she could disentangle her life from her ex-partner, William's father. In fact, it linked her life permanently to his, given that he insisted on exercising his right to share custody of William.

I thoroughly researched the relevant legislation and case law, but could not find a way in which Libby and William could be extricated from the mess.

On top of this, Libby, who was an exceptional primary school teacher, noticed that William was showing unmistakable signs of autism. Libby's assessment of William was based on her experience with schoolchildren. She had found that every primary school class that she taught included at least one or two children with autistic tendencies who were undiagnosed. She had become good at spotting them and at tailoring her teaching to their particular needs.

I researched this also, and came to the same conclusion. We were haunted by a vision of William never developing sufficiently to get a normal job, and never being able to live independently.

The fact that William was probably on the autism spectrum connected some dots for me. When he was around 9 months old, I looked after William during the day while Libby, now a single mother, went back to work as a teacher three days a week. Each morning when Libby was teaching, I would put William in the pram and take him on the train to Libby's school for breastfeeding.

William and I got on really well during the long hours we spent together. I would work on my laptop while he would spend hours on the floor doing what I called his 'physics experiments' with objects and toys. Every fifteen minutes or so, I would feel him look up towards me. I would look back at him, and we would both smile at each other. Then we would both go back happily to our important tasks.

I am now much more aware of why we bonded so quickly, deeply, and simply. We were two little autistic mates hanging out together.

As I researched autism. I was horrified to discover that early intervention was the key to overcoming some of the key limitations faced by many autistic children. Furthermore, the interventions had to involve extensive therapy—around 15 hours a week. This was very bad news because in Australia, it was extremely difficult to get a paediatrician to diagnose a child with autism before two or three years of age. This is despite the fact that interventions are initiated most productively when the child is 12 months or so.

I hit the phones. I was extremely lucky to find that the best Government-funded program in Australia for autistic children was looking for an autistic baby to participate in their program. They would diagnose William immediately. If he met their criteria, he would attend their child care center three days a week, receiving 15 hours of intensive therapy delivered by a multi-disciplinary team.

For three-year-olds, the waiting list for this program was massively over-subscribed. Parents of older children considered that to get their child into the program was like winning the lottery. But because Libby was able to notice so early that William was on the autism spectrum, there was no waiting list for him. For some time, the program had been on the lookout for autistic babies to participate in their research.

The program was a great success for William. He went from beginning to be a 'head-banger' to going to a mainstream school as a six-year-old. Now, people would be as unlikely to suspect that he is autistic than they would suspect that I am.

In the midst of these troubles, I was completely absorbed in thinking and worrying. I did not remember very often to come into the present and to watch the thinking and the worries as they arose and dissipated. Largely, I failed to use the skills that I had developed to get psychological distance from my worries.

Above all, I was focused solely on protecting my little girl Libby and her son from the world of pain that faced them. Consequently, I forgot that I would be able to deal more effectively with these challenges if I stood outside them psychologically, and was able to think about them clearly and calmly. I fell into the trap of being embedded in the problems and identified with them.

I began to notice that I was feeling continually agitated and unsettled. This was combined with feelings of danger and dread about the future. Everything I thought of doing seemed fraught with problems and impossibilities.

Quickly, this became continuous and unbearable. I did not know what was going on. I did not recognise it immediately as anxiety. Throughout my entire life, I could not remember ever feeling like this before. I had never felt anxious.

It went on and on. Nothing I did would stop it. I hardly slept. Life became impossible. On a train trip home after visiting Libby, I found that I was getting so anxious that I could not stay sitting on the train. I got off at the next station and paced around, waiting for the following train. Then, I repeated the process. It took hours to get home.

I was in no condition to research what might be causing these problems, nor to think deeply about how to use my existing skills and knowledge to overcome them. I tried a couple of doctors, a psychologist, and eventually, the 'suicide helpline', but it went on and on, 24/7.

I did not know whether anything could stop it. I thought that if I had to experience this hell for as long as I lived, I would have no option but to turn off life support. I thought that the only sensible thing to do would be to kill myself.

My last resort before this was to get booked into a mental health facility as an inpatient. I had no idea what to expect when I eventually took up this option. The only thing I had heard was that they would pump you full of Valium and possibly get you addicted to it.

The Clinic allocated me a psychiatrist. Previously, I had little contact with Jewish people. But to me, he seemed like the archetypal Jewish psychiatrist, with a big white beard, a skull cap, traditional clothes, and well over seventy years old. For me, he was a believable father figure. I found that I could put faith in his treatment. I was able to get a strong placebo effect for the first time since I was a kid. I really must have been mad.

They pumped me full of Valium together with an antidepressant and a 'mood stabilizer'. It quickly killed the anxiety. Since then, I have not had general anxiety disorder again. From time to time, I have felt anxiety in my body, but it has been intermittent, not continuous.

These transient feelings of anxiety are unlike general anxiety disorder. They do not persist indefinitely in the absence of any external events or internal circumstances that are likely to cause anxiety. The only serious anxiety I have felt since my time in the Clinic has been caused by the side effects of psychiatric drugs, or by withdrawal from them.

After two and a half weeks in the Clinic, I told my psychiatrist that I was feeling good, and that I would stop having the Valium. At the end of three weeks in the Clinic, I went home.

Unfortunately, this was not the end of my issues. But it was just the beginning of their beneficial effects on my development.

I had no withdrawal effects from ending the 20 milligrams of Valium a day that I was prescribed. I had got off it soon enough. But this was not the case when I stopped taking the 'mood stabilizer', Olanzapine. I found that when I stopped taking it, anxiety returned.

When I started taking it again, the anxiety dissipated. And so on, every time I tried to get off it.

My psychiatrist told me that the anxiety arose because the original symptoms of my mental illness were returning whenever I stopped the drug. According to him, this demonstrated that I would need to take Olanzapine for the rest of my life.

This would not have been a problem for me if Olanzapine did not produce unpleasant and debilitating side effects. The worst for me was that it disrupted my short-term memory. I found, for example, that when I was preparing a presentation to be delivered to an international conference at the Australia National University about the trajectory of evolution, I could not remember the points that I intended to make. I had to include all of these in detail on the PowerPoint slides that I used, and read from them during my talk.

Fortunately, I was now functioning well in most other respects, and I could research the scientific literature and anecdotal evidence about the issues I was experiencing with Olanzapine. I discovered that the apparent withdrawal symptoms I was suffering whenever I stopped the drug were almost certainly just that—withdrawal symptoms. They were not a recurrence of the original symptoms, and not a sign that I needed to take Olanzapine for the rest of my life.

I discovered that Olanzapine was originally developed as an antipsychotic. Because of the 'zombifying' effects that made it useful as an anti-psychotic, it could also suppress moods and changes in moods. For this reason, it has been increasingly used 'off label' as an adjunct in the treatment of mood disorders, including anxiety.

Fortunately, my research indicated that Olanzapine and other psychiatric medicines could be discontinued safely and easily if their use was tapered down over a long period of time. I applied this approach to my attempts to get off Olanzapine. After several false starts in which I reduced the dosage too quickly, I got off it. No more anxiety, no more impaired short-term memory, and no other side effects. And no return of my original illness.

However, the process of trying to wean myself off Olanzapine provided me with some extremely useful practice in enhancing my skills for becoming self-evolving. Whenever I reduced the dose, I would experience anxiety symptoms for at least a few days. They would go away if I returned to the previous dose, but doing this was not going to get me off Olanzapine once and for all.

If I was to become drug-free, and if I was to be able to conquer anxiety forever, it was obvious what I had to do. I had to come into the present, observe in detail the actual feelings in my body that I experience as anxiety, sit with those feelings without any reaction or judgment, and radically accept the feelings in my body for as long as they persisted, even if they were going to be there for the rest of my life.

Whenever I reduced my dose, I could look forward to having the opportunity to practice and strengthen these skills which are central to the development of a capacity for self-evolution. Before long, my use of the practices extinguished the psychological connection that previously existed between the feelings of anxiety and the negative reactions that the feelings triggered.

No longer would the sensations in my body that were associated with anxiety cause me to interpret the sensations as a signal that I was faced with unavoidable dangers. Now, I could still have the sensations, but they no longer operated as an unignorable alarm, signalling that I had serious issues. Now, the sensations were just sensations, no more a problem than any other kind of sensations in my body, such as the feeling of the contact between the soles of my feet and the floor.

Of course, I had known about the effectiveness of such a practice for many years. As I have outlined, I had applied it previously in limited circumstances such as when I was dieting or giving up smoking. However, my encounter with general anxiety disorder and drug withdrawal gave me the opportunity to test it in the most challenging circumstance I could imagine. If it could work with general anxiety disorder, it could work on any negative emotions or sensations. Perhaps it could work on the steps of the US Embassy in Saigon. Perhaps it could even work in hell. However, I have no desire to test those possibilities.

For the next six years, I was free from any serious anxiety. On occasions, I would become aware of feelings in my body that were like the sensations associated with anxiety. I would respond by coming into the present, examining the feelings in detail, noting whether they were caused by something that I should deal with, and accepting them radically. I would soon forget about them. This occurred less and less frequently.

In case this experience with anxiety was a sign that I was susceptible to anxiety and depression, I decided to try psychedelics. This is something that I had avoided when I was a teenager in the sixties when psychedelics were readily available. I liked the mental clarity that came with feeling sober and drug-free, and had no incentive to take any risks.

My research indicated that the psychiatric profession did not yet have any treatments that could protect me effectively against anxiety or depression. Psychiatry had been captured by the pharmaceutical corporations. The drugs it prescribed were often addictive and had harmful side effects. Psychiatrists knew very little about how to treat depression and anxiety successfully. However, there was a growing body of scientific evidence that although psychedelics were not a panacea, they could outperform standard medicines, without being addictive or having bad side effects for most people.

I commenced with an ayahuasca retreat and also, on a few occasions, tried the mushrooms that grow conveniently in Melbourne's parks and gardens each winter. I had some very interesting experiences that seemed to have possible therapeutic effects. For example, in an ayahuasca ceremony, I experienced vividly being birthed by my mother. Previously, I had always seen my mother as a tyrant who dominated and used my family for her own ends. I still do, but now I also have a strong and apparently permanent feeling of gratitude towards her.

However, my freedom from psychological problems was to end in early 2023, soon after I began writing this book.

I quickly fell into a deep depression. It was a completely different kind of experience to my bout of general anxiety disorder that I had encountered six years previously. But it was equally horrific. Both experiences were like being tortured, but in different ways. Both were life-threatening.

Similar to my earlier experience with mental illness, depression descended upon me very quickly. As before, I did not have the time to research it fully and become an expert on how to overcome it. Again, I had to rely largely on others, including the psychiatric profession, despite its ignorance and incompetence.

However, I decided that I would try to get through it myself before I would book into the mental health clinic again.

I tried, but the techniques that I had used to overcome anxiety issues did not seem to be able to be applied to depression. Anxiety manifests primarily as attention-grabbing sensations in the body. These sensations then trigger negative thinking which in turn triggers more unpleasant bodily sensations, and so on, in a vicious loop of escalating panic.

In contrast, as far as I could tell, depression did not manifest primarily as sensations in the body. Instead, it arose as extremely negative thinking that built upon itself in a vicious cycle. I had read that this cycle can often also cause and amplify anxiety. However, my techniques for dealing with anxiety apparently prevented the escalation of my depression into anxiety. Nevertheless, they were not effective against depression.

I discovered that I could escape depression temporarily by entering into the state referred to by Gurdjieff as 'self-remembering'. In this state, I would enter spacious awareness, but with part of my attention focused on my own attention, and the remainder in my environment. Placing attention on myself as well as on my environment in this way produced a powerful 'I am' experience. In this state, I was dis-embedded from any thoughts that arose, and I was free from depression.

I could get into this state while walking around the streets of Melbourne. However, intense concentration was required to enter and maintain the state. And as soon as I relaxed out of the state, the depression would return, strong as ever. Using this technique, I could avoid the depression for a time, but it did not weaken it.

My last resort before taking anti-depressants and going back to the Clinic was to try what famed psychonaut Terence McKenna referred to as a 'heroic dose' of magic mushrooms.

The heroic dose produced an extraordinary but largely indescribable experience. It catapulted me into what I experienced as different realms. They seemed to be at a higher level and more encompassing than the reality I experience here.

I felt as if I was having a series of deep insights, but I could not remember them or put them into words.

The two exceptions came as voices, not insights. The first told me that almost all the interactions, strivings, and concerns that we have during our life at this level are irrelevant at higher levels. It is meaningless noise, no more important than the interactions between atoms and molecules in a gas from our perspective.

I remember every word spoken by the second voice: "Very clever, John. But you don't know shit about what is really going on here."

I had to laugh at that.

But the trip did not act as a magic bullet. The following three days were free of depression, but then it returned in full. I was back to hell on Earth.

However, the two insights that I remembered continued to worm their way into my unconscious mind, and eventually seemed to contribute to some breakthroughs in my evolutionary thinking. But even when they did so, this did not seem to have any direct impact on my depression. Nevertheless, these experiences did help me to survive the bout of mental illness, and fundamentally changed my attitude towards suffering.

I went back to the Victoria Clinic, the same facility I had been in six years earlier. It had some great positives: The staff were extraordinarily well-trained and friendly; the Clinic had a good team of psychologists who provided sessions each day about psychological approaches to treating mental illness; and the Clinic only took in patients with manageable mental illnesses, not including psychoses. Given that the Clinic charged over a thousand dollars a day, nearly all inpatients, including myself, had private health insurance that covered the costs.

I also returned to the psychiatrist who had treated me for anxiety. However, the placebo effect had almost entirely dissipated. This was not because of any failure on his part to implement appropriate psychiatric treatments. It was because I now knew enough about psychiatry to understand that it had very little scientific knowledge about how to treat mental illness effectively.

It is not just that psychiatrists are primarily just prescribers of drugs. It is far worse than that. The drugs often do not work and, almost universally, are addictive and have harsh side effects. Largely, the standard treatment strategy is to keep trying different drugs until the patient shows some improvement (for whatever reason), or commits suicide. For some unfortunate patients, it is more like Russian roulette.

But as well as prescribing drugs for me, my psychiatrist also put a lot of additional psychotherapeutic work into me. I much appreciated these efforts and took them very seriously at the time.

Unsurprisingly, he was convinced that my depression had been precipitated by the fact that my two daughters had recently cancelled me. It seemed to be a reasonable hypothesis, but even in my depressive state, it did not ring true to me. I thought I had come to terms with my daughters' rejection. I had recognised that my older daughter Anna and I lived on different planets. We were very different kinds of human beings.

I was extremely proud of Anna when she was young because she was capable and talented in all areas, including those that I was not. In particular, she did not share my autistic tendencies at all. In fact, as well as being very smart, she was emotionally highly developed and sensitive. In contrast, I am emotionally stunted and have had to work hard on myself to develop any emotional sensibilities at all.

It took me until I was in my sixties before I finally realized that Anna's childhood might not have been as much fun and as uplifting as it was for me. In particular, I never validated her emotions. Whenever she or the family was confronted with a challenge, my go-to strategy was to think my way through to a solution. I tended to be dismissive of using emotions to guide a way to a solution. Perhaps more significantly, I did not consider that the emotional impact of outcomes was particularly important in deciding a way forward.

My rejection of the importance of emotions does not appear to be because they are suppressed in me. Rather, it seems to be because many are absent in me (of course, suppression can easily be misidentified as absence. However, the considerable and diverse efforts that I have made during my life to enable these missing emotions to surface have not been productive. It seems more likely that their continued absence is a consequence of autistic tendencies.)

In contrast, Anna often used her emotional reactions as important factors in her decision-making. While the compass I used to navigate my environment was ultra-rational, her compass seemed to combine rationality and emotions, with emotions paramount.

Consequently, throughout her upbringing, I continually dismissed and invalidated the part of her that was most central to her being. I would point out to her in laborious detail the superiority of rationality in specific circumstances. Using impeccable logic, I would demonstrate that whenever reason was dominated by emotions, it often tended to lead to ineffective decisions. To her, I would have appeared to reject and trivialize who she was in her very essence.

If I had my time over again as a parent, I would do things very differently. But I would almost certainly make other mistakes. I would also have to remember that some traumas that are unintentionally induced by parental actions might be essential for propelling the higher development of their children. It is not easy being a parent.

I wanted very much to be part of my daughters' lives and those of their children. I had spent a lot of time with the grandchildren up until my daughters were nearly forty and until the older grandkids were about to start school. For years, I had looked after one or more of the children for a full day a week, and loved every moment of it.

However, I had accepted that there is no guarantee that your children will get on well with you. Certainly, I had not felt love for my own parents, and did not get on very well at all with my mother. But after my ayahuasca experience, if she were still alive, I would waste no time apologising to her for having been a difficult and thankless child.

For these reasons, I believed that I had come to terms with my daughters' rejection of me. I barely thought about it any longer. Furthermore, there is a lot of truth in the old adage that time heals all, and I hoped that we would eventually reconcile before I died. Consequently, I was sceptical that the issue with my daughters had produced my depression. Nevertheless, I tried to take my psychiatrist's and psychologists' advice and therapy seriously.

For example, I researched the grieving process and the stages that individuals tend to go through before they eventually accept a significant loss. But the more I examined my past life, the more I realized that I had never felt grief in my life. I had not felt it when my favourite pet dog died when I was a little boy, or when my mother died, or when my wife left me, or even on the death of my father, who I liked quite a lot.

In fact, I remember that on the day we heard that my father had died, Libby was walking around the house asking anyone who cared to listen how she was supposed to feel now that her grandfather had just died. I have often told people that Libby is the person in the world whose mind works most like mine. Anna dubbed Libby my 'mini-me'. More recently, Libby herself, like her son William, has been formally diagnosed as being on the autism spectrum. I have not wasted any time wondering where she got it from.

In retrospect, I spent a lot more time in the Clinic than was healthy thinking about how my daughters' rejection had impacted me. In my heavily depressed state, my defences were down. I experienced emotions far more strongly than I did when I was not depressed. I almost became normal.

Being depressed also caused me to take more seriously than I should have other possible causes of my depression. The psychologists were very big on using mindfulness and cognitive behavioural therapy to identify and treat negative patterns of thinking. Not surprisingly, when I used these approaches, I quickly found that my thinking was riddled with self-destructive, negative thoughts.

I found that I was catastrophizing about nearly everything, including thinking that I would not survive this bout of depression. I had long known that I was a perfectionist and a control freak. More recently, I had realized that these are common symptoms of being on the autism spectrum. But now, with the input from the psychologists, I was becoming convinced that these traits might be a significant factor in the cause of my depression.

The psychologists and the social workers were very adamant about the essentiality for human flourishing of good friendships and membership in a supportive community. I had been a loner all my life, and this was amplified during the Covid pandemic. The social worker, in particular, was very concerned that when I left the Clinic, I would be returning to my single-bedroom apartment in Melbourne's Central Business District, where I lived alone. Was my depression a sign that my social isolation was finally catching up with me?

For the first six months or so after leaving the Clinic, I continued to be prescribed heavy doses of anti-depressants as well as addictive Z drugs for sleep. During this time, I tended to believe that the problems with my daughters combined with these psychological traits were the primary cause of my depression. I thought that I had to work my way through resolving these difficulties if I were to end my depression.

However, once I got off the drugs and began to feel like my old self again, these beliefs fell away.

I discovered that these apparent problems were largely symptoms, not causes, of my depression. In particular, I found that I was no longer preoccupied with my daughters to any extent. As I had before the depression, I could fully and openly accept that my girls were very different from me, and that it was unreasonable to expect that we would always be one big, happy family.

I could also now see that the catastrophizing, perfectionism, and negative thinking that plagued me while I was depressed, was not such a problem for me before I became depressed or after it lifted. I think a lot, and always have thought a lot, including about things that could go wrong. But for me, this was mostly easy and a joy. It rarely produced persistent negative feelings or sadness.

As long as I can remember, people have said to me that my life must be extremely hard work, given that I tend to think and plan and theorize about everything. I am sure that if they had to live their life the way I do, they would indeed find it exhausting and ultimately unbearable. But for me, being a perfectionist and control freak is largely effortless and enjoyable. And every day, I thank God that I can live alone in peace and quiet in my one-bedroom apartment in the center of Melbourne.

After a lot more research and experimentation, I have concluded that my mental health issues, including my bout of anxiety 6 years before, stem at least initially from a metabolic problem. It seems likely that once this metabolic issue causes mental deterioration, the decline is then amplified by my autistic tendencies. Foremost among these is my need to control my environment and the anxiety and negative thinking that arise when I fail to do so. It seems likely that such a loss of control propels me further down the path of depression and anxiety.

When I began to explore the possibility that my mental issues have a metabolic origin, I soon realized that this problem had been evident since I was very young. It is likely associated with the 'Alzheimer's gene' that I have recently discovered I carry. Broadly, my life-long symptoms are consistent with a deficiency in the provision of glucose to my brain, probably stemming from insulin resistance in the brain.

This would explain, among other things, why I have typically felt extremely tired each morning. A couple of hours after breakfast, I feel an almost irresistible urge to go back to sleep. As a first-year student at the University of Queensland, after the 9 a.m. lecture finished, I would often walk to the main library, find an empty cubicle, and go to sleep for an hour or so.

Fortunately, there is a cure for such a condition. During Covid, I had gone on the ketogenic diet for a year and had felt the best mentally that I had ever experienced. Unfortunately, I had to go off it because of the effects that it had on my blood lipid levels (I found later that this was probably due to my Alzheimer's gene. It could be overcome by minimizing my intake of saturated fats and consuming healthier fats instead).

Arguably, the keto diet had these positive effects on me because it can overcome imperfections in glucose regulation in the brain—it replaces glucose with ketones as the brain's main fuel source.

I have now returned to a modified ketogenic diet that focuses on healthier fats. I am feeling as good as I did during my first experience with the diet. And I am now medication-free.

I have also worked further on using meditation-like practices to deal with depression. If successful, these techniques will help to prevent a recurrence of depression in the event that metabolic issues again generate initial mental decline. They will also be very useful if I am wrong about the metabolic issue, and if depression arises again 'spontaneously', without a readily apparent cause.

Of course, several alternative possible explanations of my anxiety and depression seem plausible and cannot be ruled out based on any evidence. One that I keep in mind suggests that depression and anxiety serve a developmental and evolutionary function. They are irruptions from the unconscious mind that strongly incentivize us to get our development moving again if we become stuck and unable to cope with our challenges. My mental illnesses have certainly motivated me to work further on my development.

For my exploration of meditation-like practices, the techniques I developed for anxiety were a good starting point. As I mentioned earlier, I learned how to use a particular form of practice to deal with anxiety. In short, this involved accepting anxious feelings as mere sensations arising in the body, and not as signs of danger or of anything to be anxious about. Whenever the sensations trigger anxious thinking, the practice requires dis-embedding from the thoughts and letting them go by, without treating them seriously or identifying with them. This practice de-links the sensations in the body from the mental processes that would otherwise treat the sensations as an alarm that is warning of imminent threats.

This general approach was relatively easy to apply to anxiety—it is not difficult to dis-embed from and accept the bodily sensations as well as any negative thinking that they may trigger from time to time.

I began my attempts to also use this approach for depression soon after I started to get depressed. But as hard as I tried, I could not separate out in my perception the negative thinking that manifested as depression on the one hand, and the internal feelings that drove this thinking on the other. Because I was unable to distinguish and observe these internal feelings, I was unable to practice dis-embedding from them and accepting them as inert sensations.

I began searching the psychological literature and anecdotal reports in an attempt to identify anyone who had used such a strategy to deal with depression. I did not find anything of use. However, this search was hampered by the fact that I was not able to begin it before I became heavily depressed.

As was the case with anxiety, the depression started without warning, and before I knew it, I was seriously impaired. As with anxiety, I was not in a condition to undertake serious research and experimentation while I was heavily depressed, and I had no option but to endure the depression, and to rely on the mental health system and its voodoo treatments.

However, as I came out of the depression, I was able to play with practices that were somewhat analogous to the techniques that worked with anxiety.

The starting point was to come deeply into the present and accept whatever was arising in myself and my environment. Then, with a still mind, I would focus attention and awareness on the thoughts and feelings that were associated with depression. My goal was to accept radically whatever arose, and to remain as a non-attached witness to my internal processes, without getting caught up in any depressive thoughts or feelings.

What I tried to focus on observing were negative feelings that arose in my mind and then propelled depressive thinking, which in turn amplified the negative feelings, and so on, in a vicious cycle.

However, I found that the usefulness of the analogy with anxiety broke down at this point—the feelings that drive depression are not in the body, as they are with anxiety. They are not constrictions in the chest, or pains in the arm, or butterflies in the stomach. Instead, in my experience, they are unpleasant feelings that are perceived as being in the mind and that arise as part of the mental processes themselves.

The final step was to radically accept these negative feelings and to embrace them as just sensations arising within the mind. As it did for anxiety, this necessitated closely examining the sensations, locating precisely where they were in the mind, how they changed over time, and so on. Doing this would contribute to the desired outcome of experiencing the feelings as mere sensations—not as alarms that are legitimately signalling the existence of serious problems that warrant intense, depressive, negative thinking.

Ultimately, the goal was to be able to use this approach in real-time in the midst of ordinary life. The practice would involve noticing any depressive feelings that arose in my brain, acknowledging them, confirming that they were not indicative of actual problems, radically accepting them as inert sensations, and moving on.

Unfortunately, by the time that I developed this practice sufficiently, my depression had subsided. For months while I tapered off medication, I still felt mentally and physically uncomfortable, but I was no longer depressed.

I had got to the stage in my development of this practice where I could identify negative feelings as they arose in my mind and experience them as separate to my depressive thoughts. However, because my depression had largely subsided, I did not have the opportunity to test this approach fully.

What I had achieved seemed like significant progress, but I will not be able to confirm this until I have the opportunity to test it in the face of re-emerging depression. I do not want to experience again the 'hell on Earth' that characterised much of 2023 for me. But if my metabolic interventions fail me at any point, a bonus will be the opportunity to test and refine this practice.

21.

The Gifts of Autism

Another unexpected benefit of my time in the mental health facility was that I met other inmates who have autistic tendencies. We ended up forming a small group who tended to understand each other better than did the psychologists and other staff.

Our little band of mostly sisters tended to be brought closer together whenever a psychologist told a group session that humans were social organisms who needed to feel part of a supportive community to be truly happy. I would often remind the psychologist that this was not universally the case: for those who are introverted and have altruistic tendencies, hell can be other people.

Six years earlier when my little mate and grandson William had been diagnosed as being on the autism spectrum, I had researched in considerable depth the characteristics of autism. For many years previously, I had noticed that I had some autistic tendencies. Gradually, I had come to realize that I was not 'normal'. For example, as I mentioned earlier, I was nicknamed 'the odd fellow' by my boss when I was a teenager in the public service in Canberra. But I had not thought about it much. Throughout my life, I have generally been happy, including in terms of who I am. I considered myself very fortunate that I did not care much about what others thought of me or whether I had any friends.

However, deepening my understanding of autism when William was diagnosed proved to be extremely useful for me. It enabled me to connect a lot of dots about myself as well as others. I could now see that many of my personality attributes and behavioural tendencies were manifestations of autistic tendencies.

A specific discovery that my research uncovered about autism explained a lot about why I had developed my unusual cognitive capacities. Research revealed that many autistic individuals are driven by their condition to excel at building declarative models of phenomena that are of interest to them. These are the models developed by science. Many great scientists show unmistakable autistic tendencies.

A 'normal' individual comes into the world with a comprehensive set of innate mechanisms that quickly enable them to develop intuitive, procedural knowledge about how to interact with their mother, family, and other individuals. Typically, they do not have to think through in detail how to behave appropriately in many social situations. They do not have to consult mental models of their social interactions in order to identify how to act socially.

In contrast, many autistic individuals never develop this intuitive, procedural knowledge to any degree. Often, they find that engaging in small talk, shaking hands, hugging, and other social rituals seem absurd and meaningless. When they encounter complex social circumstances, they do not know intuitively what to do to fit in. They have to think through how they should behave, using declarative mental models.

Many on the spectrum over-think because they under-feel.

This is one of the great gifts of autism. Autistic children are driven from a young age to develop declarative theories and explicit mental models about social challenges that they encounter. Their inability to deal spontaneously with social circumstances provides them with a powerful and ongoing incentive to enhance their abilities to construct and use mental models. This tends to predispose them to embark on careers in which a capacity for declarative mental modelling is essential. Whenever I walk into a workplace of computer programmers, data analysts, or engineers, I see signs of autism everywhere.

The powerful incentives provided by autism for building explicit mental models has played a significant role in the evolution of humanity in the last two hundred thousand years. In particular, the shift from procedural to declarative knowledge amongst humans was driven in part by this incentivizing.

Before this transition, human behaviour was almost entirely orchestrated by procedural knowledge embodied in a wide range of skills. Some of these skills were innate, and others were learned primarily through trial-and-error associative and operant learning.

The capacity to use thinking and model building consciously to solve problems was poorly developed. The emergence of a capacity for conscious mental modelling constituted a major enhancement in evolvability. Procedural knowledge is generally limited to the learning of specific skills that are adapted to particular circumstances and contexts. But the use of mental modelling enables learning to be generalised to other circumstances. In large part this generalizability is enabled by a conscious ability to vary the parameters of a mental model and to simulate the outcomes of the changes.

Over the most recent two hundred thousand years, the ability of humans to replace procedural knowledge with declarative models has expanded progressively. This has given rise to the emergence of language—it enabled the contents of declarative modelling to be transmitted easily between individuals. This expansion of declarative modelling also eventually led to the development of a capacity to include abstractions in mental models, further enhancing their power and generalizability. As we have noted, this emergence of analytical/rational thinking powered the rise of science and technology, beginning with the First Enlightenment.

Individuals with autistic tendencies have been at the forefront of these cognitive advances. Their stunted abilities to cope intuitively with social situations and challenges provided them with strong incentives to develop alternative means for doing so.

Often, autistic children suffer from anxiety whenever their attempts to fit in socially fail. But this means that they will experience powerful psychological rewards whenever they learn how to avoid this anxiety in particular situations—for example, when they discover other methods for dealing with social and other challenges. Often, they learn that they can ward off anxiety by becoming absorbed in repetitive movements, perfectionism, obsessive focusing on details, and other means of controlling their inner and outer environments.

In 'normal' society, these kinds of strategies for overcoming anxiety can manifest as maladaptive traits. However, another strategy is possible for individuals with the potential to develop advanced model-building capacities. They can use their model-building abilities to identify how to fit in socially, at least to some extent. The potential of the challenges that autistic individuals face to reward conscious model-building in this way has played a critically important role in the recent evolution of human cognition.

My research and self-observation led to a further discovery about autism that was also very significant for understanding my own cognitive development. I was surprised to find that the potential of autistic tendencies to drive cognitive development was not limited only to 'left-brain', hyper-rational capacities.

Before this realization, I had unthinkingly adopted the common misconception that individuals with high-functioning autism tend to excel mainly in pursuits that are heavily intellectualised. I believed that high-functioning autism was great for science, engineering, and IT. But it was not so good for literary, artistic, and other highly-creative pursuits.

But then I discovered that some very prominent literary giants were on the spectrum. In the past, they were often labelled as having Asperger's Syndrome or high-functioning autism. Perhaps foremost among those who were unquestionably on the spectrum was James Joyce, who many would rank as the greatest literary figure of the 20th century.

As I investigated the issue more deeply, I realized that the forces that tended to incentivize an autistic individual to develop high-level analytical/rational thinking would not cease to operate once the individual reached that level. Instead, the incentives would continue to propel further development beyond the analytical/rational level.

In order for individuals to be able to control their environment effectively, they needed metasystemic cognition. Analytical/rational cognition enables them to build effective models of only a small proportion of their environment. But as we have discussed, most of our environment is too complex to be modelled and understood by analytical/rational cognition. This is particularly the case for social phenomena. To control their entire environment, an autistic individual needs metasystemic cognition.

These forces and patterns certainly seem to have been prominent in my life. I found that in order to achieve my goals, I had to go well beyond analytical/rational cognition. To do so required me to set out intentionally to integrate right-brain capacities with my left-brain analytical thinking. A cognitive synthesis at a higher level was required. I had to use meditation-like practices and other approaches to achieve this. The result was an ability to construct mental models of complex phenomena, enabling me to control and manipulate complex circumstances where this was possible.

James Joyce's novels appear to evidence an ability to construct mental models of complex literary challenges. The final few pages of the story 'The Dead' in his book 'Dubliners' are a wonderful example of his ability to craft combinations of words and ideas that evoke an epiphany in many readers.

It seems that Joyce set out intentionally to invent and shape the various literary devices that litter his works. His innovations did not just arise spontaneously and intuitively out of the depths of his unconscious mind. Instead, he consciously devised techniques for creating particular effects in the reader. He seems to have used this declarative understanding in combination with right-brain capacities to hone and tune the devices in order to maximize the impacts that he set out to achieve.

This propensity of autistic individuals to develop mental models of greater complexity is likely to be very significant for the spread of metasystemic cognition across humanity, and for the associated emergence of the Second Enlightenment. In particular, individuals with autistic tendencies are likely to be at the forefront of developing strategies for scaffolding metasystemic cognition, including practices that enable the integration of the 'right brain' capacities that are also necessary for higher cognition.

We will consider these scaffolding strategies and practices in detail in the book's final two chapters.

22.

The Meaning and Purpose of Life and Suffering

Despite being highly trained to focus on dispensing medications that frequently cause more harm than good, my psychiatrist did make a significant contribution to my well-being. After six or so weeks in the Clinic, I had not made any progress. Nonetheless, I told my psychiatrist that I wanted to go home. I felt that I was deteriorating physically as well as mentally in the Clinic. It seemed to me that I would be better off at home.

This seemed to worry my psychiatrist considerably, and caused him to try a different approach. I think that he was concerned that I had enough and wanted to go home and kill myself. He did not actually say anything to me along these lines. But if this is what he was thinking, he did not have it completely wrong.

During my experience with general anxiety disorder and now with depression, I had always thought that if things did not improve and did not look like they would ever improve, the sensible thing to do would be to opt-out permanently. There seemed little point in enduring the horror of serious depression or anxiety, if it was not going to end. I had always seen suicide as being on the table, but as the absolute last resort.

My psychiatrist seemed to sense this. He began to talk to me about the need for human beings to believe in something much larger than themselves that would continue to exist long after they died. This could give them meaning and purpose in their life, he suggested. My immediate reaction to this was that the evolutionary worldview provided me with abundant meaning and purpose—probably more than was experienced by most others on the planet. It provided me with a role and a purpose in a much larger scheme of things that will outlive everyone I know.

His response was yes, but did I not accept that there is an ineradicable mystery at the heart of human existence that even the evolutionary worldview cannot explain? I responded that yes, I accept this, but for me, it is a very circumscribed mystery—once the big bang occurred, there was no mystery as to how the universe unfolded: known physical laws and material processes eventually gave rise to life; this life increased progressively in hierarchical complexity and evolvability; evolution eventually gave rise to a cognitive capacity that could comprehend the processes that produced this evolutionary trajectory; ultimately this enabled a shift to intentional evolution; and so on, and so on.

But I had long recognised that none of this explains why there is a universe in the first place. I had accepted that this mystery was impenetrable. Evidence did not exist that enabled science to peer beyond the veil and to develop scientific explanations of why there is something rather than nothing.

I knew that it was possible to construct an infinite number of hypotheses that each could explain logically why the universe exists and that cannot be contradicted by any known facts. For example, all sorts of gods with all sorts of powers could be hypothesised, as could multitudes of possible material causes. However, there is no evidence that could be used to falsify any of these hypotheses or to narrow them down to a single hypothesis.

Furthermore, if some means were found to enable science to develop an understanding of what generated the Big Bang, this would push back the horizon of our knowledge somewhat, but there would still be a horizon beyond which we could not know anything with certainty.

Consequently, I agreed with my psychiatrist that there was an ineradicable mystery at the heart of human existence and, in fact, at the heart of any existence at all. However, at first, I did not see that this had any significant implications.

But as his questions wormed their way into my mind, they connected with other realizations and intuitions. In particular, they resonated with the messages I received from the voices I heard during my mushroom experience: there are larger-scale processes at much higher levels than ours, and I do not know much about what is really going on here.

Previously, I had rejected spending much time thinking about what might have produced our universe in the first place. I believed that doing so was a dead end. It was impossible to make any progress in sorting through the endless possibilities.

However, I began to realize that I had been using ineffective methods to address these issues. I had been proceeding as if my primary goal was to develop an objective and scientific explanation of the causes of the universe. But now I began to understand that I should have seen this approach as just one of several possible means for discovering what I really wanted to know, not just an end in itself.

In fact, my central interest was whether the nature of these causal processes might have implications for us, here and now. My main goal was to discover whether an understanding of the processes could answer questions like: Was our universe established for a particular purpose? What role, if any, do we have in fulfilling this purpose? Will it make any difference to us whether we consciously set out to do what we can to assist the achievement of this purpose?

Of course, if a scientific approach could explain in detail what caused our universe to emerge, it would greatly facilitate the answering of these and other questions about the sense and significance of our existence. Once science understands these causes sufficiently, its discoveries can be used to work out whether there are any implications for the way we live our lives.

However, we can probably never know with scientific certainty the precise nature of the processes that produced our universe. We are unlikely ever to be certain about whether these unknown processes have major implications for our lives.

In the absence of a scientific explanation, the possibilities are endless. Depending on the nature of the possibilities, what we do during our lives might make no difference. For example, there may be no particular way in which we can act that will pay off for us in the longer term, e.g. by enabling us to survive after death. But neither can we rule out possibilities that would provide such payoffs. We can imagine

293

numerous plausible possibilities that have the potential to provide meaning and purpose for at least some forms of human existence.

Nonetheless, in the face of this uncertainty, are there rational methods we can use to decide how we should live our lives now to maximize our longer-term interests? Increasingly, I realized that this was the most relevant question to ask about the fundamental nature of our reality.

Even if there is radical uncertainty about why there is something rather than nothing, is it still possible to make rational decisions about how we should act now? Is there a method for making choices that does not rely on certain knowledge, but that, for example, maximizes our chances of surviving in the long term, including after our bodies die?

Of course, humans make decisions regularly in the face of uncertainty. Techniques have been developed and tested that enable them to do so rationally and effectively, even in the face of radical uncertainty. Radical uncertainty exists when it is impossible to predict accurately the future consequences of possible actions, or even to assign probabilities to the possible outcomes. Humans face radical uncertainty about the reasons for the existence of our universe and their implications for our lives.

These decision-making methods can be particularly effective when at least some of the possible outcomes of decisions can produce significant benefits for the decision-maker. As we shall see, this is the case even though there may be many more possible outcomes that do not impact on the decision-maker at all.

An infinite number of possibilities exist that are consistent with all available evidence that could explain the existence of our universe. Some of these possibilities have the potential to make considerable sense of our universe in general and of human existence in particular. Significantly, these meaningful possibilities cannot be ruled out. They are actual possibilities that seem plausible and that are not falsified by any known evidence.

I began to realize that it is possible, for example, that the larger-scale processes that are responsible for the existence of our universe may have set it up for the purpose of generating higher intelligences. The path that our universe provides for the development of these capacities might include accepting suffering and learning as much as possible from enduring it. If this were the case, and it cannot be ruled out, the option of turning off life support might not be as rationally attractive as I thought previously. It may involve closing the door to further possibilities that are valued and even rewarded by the processes that established this universe.

This perspective might be seen to differ from my original evolutionary worldview in only minor ways. However, it made a huge difference to how I would come to view my life and whether I should persevere with it even in the face of great on-going suffering.

I will now set out to identify in detail the rational decision-making methods that are capable of identifying optimal strategies despite radical uncertainty. I will then apply them to these fundamental existential issues.

But such strategies are only valid if there are plausible possibilities that cannot be ruled out and that have longer-term implications for what humans do here and now in the world. I will begin by demonstrating the existence of a number of such significant possibilities.

Among possible explanations for why our universe exists are those that suggest it was brought into existence by a being(s) that exist outside our universe and have the capacity to create universes. Their methods of creation might include producing simulations.

Why would such a powerful being(s) create our universe? It is possible to divide the possibilities into several relevant, overlapping categories. First, there are an infinite number of possible explanations that make no sense in the context of human intentions and goals. Furthermore, an infinite number of possibilities exist that have no implications for how we live our lives. No matter what we do in such a universe, our actions would have no consequences for us beyond our temporary existence. There are also an infinite number and variety of 'God hypotheses'.

However, there is a class of somewhat plausible possibilities in which the being(s) who brought our universe into existence have done so for their own particular purposes. When they established our universe, they set it up in such a way that it performs particular functions that serve their goals. But of course, it might not have been set up at all to serve human goals. No matter how we live our lives in such a universe, it might make no difference to whether we can, for example, survive death.

For example, our universe might be a simulation established by an advanced civilization that exists outside our universe. In this hypothetical example, they have established the simulation for the purpose of evaluating the consequences of alternative ways of organising their societies. It may be that because of computational irreducibility, even such advanced intelligences cannot work out the consequences for their own civilization of these alternative strategies. Consequently, they need to simulate various possibilities if they are to identify the best strategy.

This advanced civilization may be completely unconcerned that their simulated universes will produce extensive misery and suffering for the conscious beings that will emerge in the simulations.

But there is a class of possibilities that is more relevant and interesting, at least to humans. In these cases, the way we live our lives does have consequences for the possibility of some kind of existence that extends beyond our bodily deaths.

I will briefly outline three classes of this kind of possibility that cannot be ruled out.

The first class of possibilities is the creation by an advanced civilization of simulations that are designed specifically to provide intelligences with environments and experiences that enhance their psychological growth and development.

The plausibility of these possibilities is illustrated by the fact that humanity is already beginning to head in this general direction.

Human therapeutic psychology has often used visualization, imagination, and environmental manipulation to heal individuals from trauma and to build greater psychological resilience and functioning. In recent years, simulations and virtual realities have also begun to be used to deliver these kinds of therapies, albeit in a limited way. A very simple example is the use of virtual environments to extinguish phobias associated with airplane travel. The virtual environment gradually introduces the individual to the experience of being a passenger on a plane. It exposes them progressively to the negative feelings triggered by their phobia, but in an environment that keeps them calm and relaxed. Once they get used to a particular level of exposure, it is increased further, eventually extinguishing the phobia.

However, the appropriate use of complex and realistic virtual environments is likely to take this to an entirely new level. In principle, individuals could be temporarily embedded in whatever social and other environmental circumstances will produce the desired psychological outcomes.

An appropriate set of experiences could significantly change an individual's conditioning. Or it could operate at a meta-level, propelling the development of a capacity to be self-evolving. This would almost certainly have to include experiences that are traumatic for the individual, thereby motivating them to leave the comfort of their existing level of psychological development and to do the hard work on themselves that is needed to move them vertically to the next level.

Of course, in order to have maximum effect when an individual is embedded in such a virtual reality, it may be desirable to induce the individual into a state in which they believe that their virtual experiences are real. When this is achieved, the participating individual would experience the virtual therapeutic experience as being indistinguishable from real life.

This kind of technology might become very significant if humans achieve a state of abundance. Many dream that eventually, human technological capabilities, perhaps aided by AI, will be able to effortlessly meet all human needs and wants. If such a society were ever achieved, children might need to spend considerable time in a virtual reality that provides them with the shocks, traumas, and other complex challenges that are essential for successful human psychological development.

Our universe could be such a simulation created by an advanced civilization.

The second class of possibilities that I will consider is the creation by an advanced civilization of simulations that are designed to develop higher intelligences in a way that is safe for the civilization.

Again, human AI developers are already beginning to consider the desirability of such an approach. This is because AI might emerge eventually that has the power, motivation, and intelligence to harm its creators. Some believe that it may even destroy human civilization.

A solution might be to confine the training and development of AI to simulated environments that provide no opportunity for the AI to interact with the creators or their universe. These simulated 'sandbox' universes would be specifically structured so that they facilitate the development of AI of higher intelligence. Once particular intelligences reach the desired level, they could be harvested by the advanced civilization and used for their purposes.

However, harvesting would proceed only once the AI was assessed as being safe from the perspective of the advanced civilization. Additional security could be provided by embedding the simulation in a nested hierarchy of simulated environments. Escape from lower-level simulations would not threaten the civilization.

From the perspective of an advanced civilization, we might be AI that is evolving and developing in a sandbox universe that it has simulated.

The third class of possibilities arises because it may be impossible to produce highly intelligent AI by a combination of engineering and training. The only viable way to produce such AI might be to set up a simulation of an entire evolutionary process. This process would begin with the emergence of a universe. Eventually, life would emerge within the simulation and evolve in the direction of increasing complexity and evolvability. If it were set up appropriately, such a simulation would eventually produce high-level intelligences.

This class of possibilities cannot yet be ruled out. This is despite the fact that many AI researchers believe strongly that it is possible to engineer and train Artificial General Intelligences (AGI). However, none have done so, or have come even close. Human attempts to produce AGI or even to understand how it might function have not yet gone beyond wishful thinking.

At present, we know of only one particular strategy that can reliably produce human-level intelligence in an entity that begins far below this level. The strategy's starting point is a newborn baby. Typically, the baby cannot speak, think abstractly, coordinate its bodily movements, or undertake the other myriad functions that characterise human-level capacities.

Thirty years later, and with a bit of luck, the typical baby will have grown into an adult with human-level intelligence. In the early years, the physical maturation of the brain and other bodily systems will have contributed to this development. But mostly, the intelligence of the growing human will develop due to learning and associated processes that help to install the relevant capabilities.

What kinds of experiences, interactions, and training does a growing human need if its intelligence is to develop appropriately? Could an AI that begins with baby-level intelligence be subjected to similar experiences, interactions, and training, thereby achieving comparable improvements in intelligence? Furthermore, are there alternative approaches that could be taken with AI that could short-circuit our current methods of producing human-level intelligence? For example, to what extent could AI be programmed directly with capacities that humans have to learn?

The current state of human knowledge about how to develop human-level intelligence in humans is very limited. It happens right in front of us, but we actually know very little about how to do it in humans, let alone in AI. As I have outlined, even I made one or two mistakes while raising my daughters.

However, it is possible to get a general idea of what is necessary to produce human-level intelligence in humans. This can be achieved by identifying in broad terms the kinds of learning experiences that children are subjected to as they develop. We can get a sense of the learning experiences that are most important by noting those that are essential for the successful development of the child.

HUMAN SUPERINTELLIGENCE

It is well known that the foundations of our cognitive and social-emotional development are established in the first year or so of life through long sequences of interactions with our mothers or other primary caregivers. This then broadens into the complex processes of socialization that generally involve other family members and then wider communities. Again, this comprises innumerable interactions and learning experiences. We have little idea about which particular interactions are critical for our successful development. But we know, in general, that these kinds of complex interactions and learning experiences are essential if we are to continue to grow and develop successfully.

In a modern complex society, this is followed by immersion in pre-school and kindergarten, and then 12 long years of schooling. Without something like this, we will not develop the cognitive and social/emotional knowledge and capacities to get well-paid work in a modern economic system.

But formal education is not the only critical factor during these years. Play and other complex interactions with peers that occur before and during school are also significant in developing all kinds of social/emotional and cognitive skills.

After schooling comes university or work and entanglement in a wider community. Again, myriads of interactions and learning experiences occur. It is not possible to identify the specific sequences and networks of experiences that eventually prove essential for enabling a particular career.

Whatever career we eventually find ourselves suited to, it may enable us to make a significant contribution to the complex society in which we live and work. However, all the intervening steps that took us there could not have been planned in advance. Many skills and capacities that prove significant in our lives are acquired due to chance meetings with individuals who grew up in completely different environments or even in different countries and cultures.

Failures, traumas, abusive relationships (including within the family), and mental illness may, in retrospect, have been critically important for driving our acquisition of skills and knowledge that eventually prove essential for our development. Would we develop fully as intelligent human beings without encountering challenges that cause us to, for example, engage in self-reflection; experience failed friendships; fall in and out of love; learn from bitter experiences that our behaviour when younger was inappropriate; realize many years after this that the revised behaviours we adopted were also inappropriate; move through Piaget's levels of cognitive development; learn how to meditate to accelerate this developmental process; realize that an unexamined life is not worth living; develop psychological defences and then modify them and eventually drop them; and so on, and so on?

How could we possibly provide AI with these experiences that appear essential for developing intelligence capable of achieving complex goals in our world? We would not know where to start.

At present, humans have little knowledge about which kinds of experiences might be essential to develop our cognitive and social-emotional capacities. As this book outlines in relation to my life, the importance of these experiences might not be at all obvious as they occur. As Soren Kierkegaard said, "Life can only be understood backwards, but it must be lived forwards."

In some tribal societies, it was said that it takes a village to raise a child. Now, it tends to take at least a modern nation-state. Increasingly, it is taking an international system of markets and economic activities. Eventually, it will take a living global entity.

Is this what we must do to produce human-level AI and beyond? Like us, will AI have to grow and develop in a diverse and complex society if it is ever to acquire human-level intelligence?

Do we have to begin with AI that is roughly at the same level as a human baby, and then embed it in an environment that will deliver it the myriads of learning experiences and interactions in the 'correct' sequences that eventually transform human babies into fully-functioning adults?

If so, it seems impossible to do this by designing each of these experiences and then delivering them to the AI in a training environment. Attempts to do so would be likely to be continually beset by computational irreducibilities.

It should also be obvious that this cannot be achieved by providing the AI with all human declarative knowledge that is available on the Internet and elsewhere. This will go nowhere near providing the developing AI with the myriads of complex social and physical interactions that humans experience as they develop.

Instead, it seems that the only way to proceed successfully would be to embed an agentic and goal-directed AI into a complex social environment that self-organises the necessary experiences, in interaction with the AI.

Given our current knowledge and experience, the only feasible way of doing this would be to raise the AI 'baby' as if it were a human baby. At present, the only way we could have any hope of ensuring that a developing, agentic AI will have learning experiences similar to those that can transform a human baby into a functioning adult would be to immerse it in the only environment that can currently provide such opportunities—a human society.

A similar approach was taken to test whether primates could attain human-level capacities if they were provided with appropriate learning opportunities. Primate babies were brought up in human families. However, they did not progress far.

But even if such a strategy were feasible, before we can begin to implement it, we encounter an even bigger challenge. We may not even be able to get to the starting point that I have assumed in the discussion so far. We have little idea about how to design and engineer AI that is as capable as a newly born human baby and, more importantly, that has the potential of a human baby.

This is a major obstacle. The only process that we know of that can produce anything like a human baby from scratch is the entire evolutionary process itself.

The reasons why this challenge may never be overcome are similar to the reasons why it is unlikely that we can ever produce human-level AI directly by engineering and training: much of the knowledge embodied in a human baby is procedural; it has been discovered by billions of years of evolutionary trial-and-error; computational irreducibilities make it impossible to build models that would enable us to understand and manipulate the relevant processes; as a consequence, we have no idea how to emulate what the evolutionary process has achieved or even to identify which particular steps were critical in moving towards baby-level intelligence accompanied by the potential to develop human-level intelligence; and so on.

However, if we start a universe with a Big Bang that is appropriately fine-tuned, it seems likely that it will eventually produce life, and the evolution of this life will have a trajectory. As we have discussed, the details will differ, but eventually, the trajectory seems likely to produce organisms that have human-level intelligence.

If humanity develops the capacity to simulate the emergence and evolution of such a universe, it seems likely that eventually, we would be able to harvest intelligences at the human level and beyond that develop within the universe.

I have only sketched the relevant arguments and considerations here. But it may be the case that the only way in which an advanced civilization could produce human-level AI and beyond is by simulating the formation and evolution of a universe. Such a simulation would have to be set up with initial conditions that ensure that the evolving universe is both life-friendly and intelligence-friendly and that produces an evolutionary trajectory that heads in the direction of generating life and intelligence of increasing complexity and evolvability.

Of course, if humans decide to proceed down such a path, it would highlight the possibility that our universe and our lives are the product of such a simulation. The argument made by philosopher Nick Bostrom about the probability that we are actually living in such a simulation applies equally here.[23] If producing such a simulation is something that civilizations are likely to do once they become sufficiently advanced, we are probably in such a simulation. The probability that any given civilization is the original one that gave rise to a long sequence of simulations is very small. It is much more likely that it is a simulated one.

---

[23] Bostrom (2003) – see References

But this brings me back to the central issue that propelled my thinking about these issues. To what extent can the existence of these classes of possibilities help address the ineradicable mystery that exists at the heart of human existence? They are just possibilities, incapable of eradicating the mystery completely. No one can explain with certainty why something exists rather than nothing. Why have I gone down this path knowing that it is incapable of producing certainty about the big existential question that we all face?

At best, as we have seen, these possibilities represent hypotheses that cannot be ruled out on the basis of current evidence. However, there are an infinite number of hypotheses that can explain current circumstances and are consistent with all known facts (these include an infinite number that rely on various gods and supernatural beings as well as materialistic theories). Furthermore, most of these possibilities fail to get beyond the infinite regress that arises when we ask the questions: Who or what created those being(s) that are hypothesised to have created our universe, and who or what created them, and so on, and so on, indefinitely?

However, as I suggested earlier and will now demonstrate in more detail, possibilities that cannot be established or ruled out by current evidence can provide good reasons for a rational agent to act in particular ways. This is the case even when there is no evidential basis to establish any particular possibility, or even to assign probabilities to any possibilities. Radical uncertainty of this kind is a common challenge for humans and other agents, and it is not the end of the story.

An example of such a challenge is when an agent is faced with many hypotheses that make predictions about future events. In this example, no evidence establishes any of these hypotheses or even enables probabilities to be assigned. However, consider a hypothesis that makes specific predictions about a particular future event. The predictions are such that if the agent uses the predictions to decide what actions it will take, and if the hypothesis proves to be accurate, the agent will reap considerable benefits.

A specific example is the hypothesis that our universe is a simulation created by an advanced civilization that exists outside our universe. The hypothesis postulates that the civilization created the simulation in order to evolve and develop higher forms of intelligence, safely. The advanced civilization intends to harvest suitable higher intelligences from the simulation once they emerge. These intelligences will escape physical death within the simulated universe and be deployed to perform useful functions in the advanced civilization. Consider an individual agent within the simulation who believes that it would be in its interests to be harvested either as an individual or as part of a larger-scale collective intelligence.

In this example, we will assume that all alternative hypotheses facing the agent are non-beneficial—they will not provide net benefits to the agent if they are acted upon and prove to be correct. In such a case, the agent will maximize its interests if it decides to act on the basis of the beneficial hypothesis and spends its life working on itself in order to enhance its intelligence.

Obviously, if the beneficial hypothesis proves to be correct, the agent will be substantially advantaged, continuing to live beyond physical death. Alternatively, it may happen that one of the non-beneficial hypotheses proves to be true. But even if this is the case, the agent will not end up worse off by having acted on the beneficial hypothesis that has proven to be incorrect, provided the costs of doing so are not significant.

Under these circumstances, despite the agent being faced with radical uncertainty, there is a rational strategy for deciding its actions. This strategy will maximise the achievement of its goals.

The agent does not know in advance which hypothesis is true. But nonetheless, it can decide rationally which hypothesis (or class of hypotheses) it should act upon, as if they were true.

There are many variations on this theme. For example, an agent might face a mixture of plausible hypotheses that produce different combinations of benefits and harms. The field of decision theory studies these more complex cases. It sets out to identify what a rational agent should decide in various circumstances, given their particular goals, even in the face of radical uncertainty.

Different agents might pursue dissimilar goals. The decision-making strategies that are optimal may vary for different goals. For example, decision theory evaluates what a rational agent should decide if its goal is to maximize its benefits, or to minimize harm to itself, or to balance risks and benefits in some defined way, or to minimize the maximum regret that it might experience, and so on.[24]

The existence of decision theory and its rigorous insights enable us to go beyond the reach of current science. The application of its discoveries enables us to make rational decisions about how we should live our lives, even though we face radical existential uncertainty.

What specific implications do these methods have for deciding how we should act now, given possible explanations for the existence of the universe? Fortunately, we do not have to deal with an infinitely huge number and diversity of classes of plausible hypotheses. This is because evidence is available that reduces the number of hypotheses that we need to consider. This evidence significantly decreases the number of hypotheses that can be considered to be plausible.

In particular, it is evident that we live in a universe that is consistent with the hypothesis that it is fine-tuned in many ways to be life-friendly. Furthermore, the fine-tuning is not conducive only to the emergence of simple life. It also appears to give rise to an evolutionary process with a particular trajectory. As we have seen, this trajectory results increasingly in the integration of living processes and in enhancing their evolvability/intelligence.

As the trajectory unfolds, cooperative organisations of greater and greater scale emerge, and as the scale of these increases, so too does their evolvability/intelligence. If this trajectory continues to unfold successfully on Earth, the intelligence of the emerging global entity will far surpass the intelligence of any individual human. Nonetheless, individuals will contribute to the intelligence of the global entity, just as our brain cells contribute to our intelligence.

Many possible hypotheses about why our universe exists are not plausibly consistent with its apparent fine-tuning. We can exclude them

---

[24] e.g., see Bejleri *et al* (2022) – see References for full citation

from further consideration. In general, this includes all 'right-hand path' spiritual and religious traditions that reject the enhancement of agency as a central goal of their practices and beliefs.

Of the remaining hypotheses that retain their plausibility, there are some that would provide benefits to individuals if they act in particular ways during their lives. I have already briefly discussed examples: our universe might have been intentionally set up to produce higher levels of intelligence. It may be that simulating such a universe is the only way to grow higher intelligences safely. It is possible that intelligences that emerge within such a universe will be advantaged if they contribute to the further development of themselves and of other intelligences in their universe.

For example, they may be part of intelligent processes that are subsequently harvested and used for functions outside the simulated universe, or they may perform permanent functions within the simulated universe. Alternatively, an advanced civilization might have set up a fine-tuned simulation in order to provide an optimal environment for the further development of individuals from within the civilization itself.

But the evidence that suggests our universe might have been intentionally fine-tuned also leaves us with many other hypotheses. These include ones that would not provide any existential benefits to individuals who emerge within such a universe, no matter what they do or contribute. For example, the creators of the simulation might have produced it simply to test hypotheses of their own about how living processes evolve. When their experiments reach an end, they may just discard the simulation and anything that emerged within it.

A further example is the 'multiverse' hypothesis. It explains fine-tuning by assuming that those universes in which life emerges will necessarily exhibit characteristics that enable life to emerge and develop. From the perspective of intelligent life that emerges in such a universe, it will appear to be fine-tuned for life.

Nevertheless, even when an agent cannot tell which of these kinds of universes it lives in, it can be rational for the agent to act 'As If' the predictions made by the most beneficial hypothesis are true, all other things being equal.

These broad considerations provide strong and rational reasons for the adoption of life-goals that are pro-evolutionary. This would involve doing what you can to advance the trajectory of evolution by developing your own intelligence/evolvability, as well as contributing to the development of the intelligence/evolvability of the larger-scale cooperatives in which you are embedded. On this planet at this time, a priority is to contribute to the emergence of a cooperative and highly-evolvable global society.

Of course, there are no absolute guarantees that embracing an evolutionary worldview will pay off for individuals in the longer term. But there are no downsides associated with doing so. If fact, there are many benefits: as you work on yourself to enhance your evolvability, you will no longer be harmed by negative emotions and feelings and will be able to experience satisfaction and joy in whatever you choose to do. You will also develop enhanced cognitive capacities that will enable you to generate effective strategies for achieving your goals, whatever they may be.

In summary, if you wish to live your life in a way that maximizes your existential interests, you should spend your time contributing positively to the purposes for which our universe might well have been set up. If this proves to be futile, as it may, it seems that there are no plausible alternative strategies that are likely to produce a better outcome for you.

However, I have only sketched a broad outline of the relevant considerations here. This is not the place to present a far more comprehensive model of the strategies that a rational agent should adopt in the face of radical uncertainty about why there is something rather than nothing.

Furthermore, like science itself, the conclusions reached by this kind of approach are likely to change as assessments of plausibility evolve and as more evidence emerges. New evidence might open new plausible hypotheses. Other kinds of fresh evidence might rule out some of the hypotheses that are now considered to be plausible and consistent with current evidence.

As far as I know, only one thinker has used this kind of approach previously in a serious attempt to address the big existential questions. French mathematician and philosopher Blaise Pascal used a similar line of reasoning to argue that a rational individual should act as if the Christian god actually exists, and follow the tenets of Christianity. If the individual does so, and if there actually is such a god, the individual will go to heaven and avoid hell. If the alternative proves true, the individual will lose little.

However, Pascal's approach has several obvious flaws. Foremost among these is that the only beneficial hypothesis that he considered plausible was that the Christian God existed. He did not include in his analysis any of the other religious traditions that are at least as plausible as his version of Christianity (and are equally implausible in the context of modern knowledge).

Broadly, this kind of thinking led me to decide that I will endeavour to endure and learn from any suffering that comes my way in this universe. As has been the case for many of the misfortunes in my life up until now, this suffering may provide important learning experiences and motivation for my future development.

Furthermore, if I can develop my capacity to accept fully whatever circumstances arise, I will become more like the Buddhist monk on the steps of the United States embassy in Vietnam. If I get there, there will not be any downside: whatever suffering comes my way, I will not flinch or react negatively. The experience will be no different from any other set of bodily sensations. Furthermore, as I have found, once suffering has passed, it is as if it never happened.

Of course, the idea that life in this universe might be a temporary experience that is intended to provide learning experiences is common across the great spiritual and religious traditions, as well as in New Age 'thinking'. These sources variously suggest that we are reincarnated until we develop sufficiently; that we are 'spiritual beings having a human experience'; that we are 'the one's' way of experiencing a greater range of possibilities; and so on. But as with most beliefs within the spiritual and religious traditions, their explanations and justifications are often mutually contradictory. Their beliefs need to be shorn of their spiritual mumbo-jumbo, and understood instead using rational approaches.

These realizations had a major impact on my attitudes to life and its challenges. Previously, I had fallen into the trap of concluding that if science was unable to understand why our reality exists, it was pointless to consider the issues any further.

This led me to conclude that any thinking about possible explanations for existence was senseless and futile and could not possibly go anywhere. Science was our best method for working out how the world works, and if things were beyond the reach of science, they were beyond rational knowing.

But this thinking prevented me from seeing the implications of another obvious possibility. Although we do not have the science-based tools to account for why our reality exists and probably never will have those tools, this does not bring our explorations to an end. It does not prove that there is no possible explanation for our existence that can make sense of our lives. It suggests only that if there is such an explanation, we are incapable of demonstrating its validity during our lives. But all this means is that we cannot rule out any particular possibility. For example, it does not enable us to rule out the possibility that when our bodies die, our intelligence and being may continue to exist in some other realm. Neither can we rule out the alternative.

We simply do not know.

We cannot know what death will bring us. But this means that we will have no reason to be surprised if one of the more 'hopeful' possibilities proves to be true. Consequently, the absence of certainty about these possibilities should not determine how we enter the dying process. It is consistent with all scientific knowledge and other facts to enter the dying process with unrestricted curiosity about what, if anything, comes next. As the voice told me when I took my heroic dose of mushrooms: "Very clever, John. But you don't know shit about what is really going on here." I now take this message to be true, but in a positive sense. What is really going on here might make additional sense of my existence.

Furthermore, as I have outlined, there are ways in which we can live here and now that will pay off in particular kinds of universes—for example, in universes that provide us with continued existence if we work on developing ourselves and the larger living systems in which we are embedded. Broadly, this entails aligning our lives and actions with the trajectory of evolution. The evolutionary worldview identifies in detail what this requires.

These considerations led me to a conclusion that was not very different from the position I had reached before my bout of depression. But the context in which I reached it has changed radically. I am now very much open to the possibility that what is going on here in our universe makes sense in some larger context that includes but transcends this universe.

At the end of all my explorations, I arrived back where I had started, but understood its wider implications for the first time.

As I have indicated, this has fundamentally changed my attitude to suffering and death. I still do not know with any certainty what is really going on here. But now I know that it is possible to make rational decisions about things beyond the reach of science and about which I have no certain knowledge.

Ultimately, it may prove existentially futile to live my life as if the values that seem to be implicit in our universe also make sense in a larger scheme of things. But amongst the many possible hypotheses that could account for why there is something rather than nothing, the pay-off if such a hypothesis turns out to be correct is potentially very significant.

* * *

However, I was unsettled by the conclusion that this new kind of rational thinking could deal effectively with issues that apparently fall outside the reach of science.

Up until this point, my thinking had proceeded on the assumption that scientific investigation, broadly understood, is humanity's best method for discovering how we should act in the world to achieve our goals, whatever they may be. However, I had now developed a new, rational method for deciding how to act in circumstances that could not be understood using scientific methods. This new approach seemed to be a reliable and rational decision-making method.

Was it possible to unite the two forms of thinking in a new synthesis? In such a synthesis, each of these forms of thinking would be seen as instances of a single, higher-level framework. The new framework would embody a unified approach that subsumed both of these forms of thinking. It would identify rational methods for making decisions that worked effectively in both domains—for science and for circumstances characterised by radical uncertainty.

The potential for such a possibility seemed to me to be heightened by the fact that the methodology of science has never been properly established. Science has never demonstrated from first principles how its methods overcome the problem of induction and the associated problem of causation. Nor has philosophy.

The great Scottish philosopher David Hume demonstrated the seriousness of these problems. Broadly, he pointed out that when a particular event A occurs, and when event B has always been found to follow event A, this does not prove that A causes B or even that the next time A occurs, B will occur again. In these circumstances, it does not even prove that B will occur with greater probability.

What Hume argued is incontrovertible. No matter how many times event A is followed by event B, it does not rule out the possibility that the next time A occurs, B will not.

Science survived this devastating attack on its foundations because of its apparent success at discovering laws of nature and harnessing them to satisfy human goals. There is abundant evidence that science has worked brilliantly in practice, despite Hume's arguments.

Science's response to Hume has not been to disprove his arguments. Instead, it has been to just get on with science and continue producing extremely valuable discoveries.

Scientific methods have been claimed to discover many regularities and laws. Generally, these regularities continue to manifest whenever their existence is tested by observations. But of course, this

312

does not mean that the apparent past regularities will hold true in the next minute or hour. Furthermore, as Hume pointed out, the fact that scientific methods appear to have worked to some extent in the past, does not mean that they will continue to work in the future.

Nor is it the case that the more times that an event B occurs immediately after A occurs, the more probable it is that B will be found to follow A in the future. Probabilities cannot be rationally assigned to future events on the basis of past experiences. Hume's argument applies equally to assigning probabilities as it does to predicting certainty.

In any specific case, it is impossible to rule out the possibility that our universe might be arranged in such a way that in the next instant, all previous apparent laws of nature will cease to apply. Or the possibility that any one, or any combination of them, will cease to apply.

The possibility cannot be ruled out that apparent laws may be replaced at any instant by any one of an infinite number of alternative laws that would produce entirely different outcomes. From this instant onwards, this may occur every second, or millisecond. There are endless possibilities. None of them are ruled out by the observation that up until now, the rules of nature that we believe to have discovered in the past, appear to have continued to apply.

Furthermore, there is nothing that enables us to say with any validity that any new laws that emerge will continue to apply in the future. Nor can we demonstrate with any certainty that the longer a new law applies, the more probable it is that it will continue to apply. And so on, and so on.

Karl Popper's ideas about the philosophy and methods of science are perhaps the most widely accepted within science. He agreed with Hume that observations that are consistent with the predictions of a scientific hypothesis can never demonstrate that the hypothesis is correct.

However, Popper argued that observations that are inconsistent with the predictions of a hypothesis can falsify or disconfirm it. He argued that contrary to naïve inductionism, science progresses not through the verification and confirmation of hypotheses, but by the refutation of 'wrong' hypotheses. Logically, hypotheses can be disproven, but never proven. According to Popper, science progresses through the accumulation of hypotheses that have survived all attempts to falsify them, and through the rejection of falsified hypotheses.

But did Popper's approach really overcome Hume's arguments? Did it actually put science on sound foundations?

No, it did not. As Hume might point out, any number of failed attempts to falsify a hypothesis does not prove that in the next instant in time, new attempts to falsify it will also fail. The predictions of hypotheses that were falsified in the past may make predictions that prove correct at any instant in the future. This and an infinite number of other possibilities cannot be ruled out in each new instant, no matter what possibilities actually occurred previously. At every future instant, an infinite number of possible hypotheses come back into possibility, including those that have been falsified previously.

Is it possible to find some other way in which science can be placed on sound foundations? The answer to this fundamental question depends on what we take to be meant by 'sound foundations'. If we are asking whether science can arrive at certain knowledge about the future, the answer is clearly no. Throughout history, science and philosophy have stubbornly sought a scientific methodology that can be proven to be capable of discovering truths that will apply in the future. But this pursuit has failed. Science and philosophy have been relentlessly exploring a blind alley.

There can be no such proof. Hume's arguments demonstrate this. Science and philosophy have been asking the wrong questions.

The way through these difficulties is broadly the same as I developed for dealing with the problem of existence. It is founded on the realization that an agent can act in ways that qualify as rational, even when confronted with radical uncertainty. Here, rational decisions are taken to be those that can be demonstrated to increase the likelihood that the agent will achieve its goals. Radical uncertainty is faced by an agent in relation to a particular decision when forecasting is impossible, and the agent is unable even to assign probabilities to the possible outcomes of its actions.

As we have seen in relation to existential uncertainty, even when facing radical uncertainty, an agent can still arrive at a decision using a method that will maximize the achievement of its interests, provided particular conditions are met.

The example of existential uncertainty that we explored considered circumstances in which a particular subset of possible outcomes existed amongst a much larger set of possibilities. For this subset, if the agent acts in a particular way, and if this possible outcome actually occurs, the agent will benefit. Provided that the downsides experienced by the agent are small when it acts in this way, and there are no other possible outcomes that would benefit the agent, it is in its rational interests to decide to act in this particular way. Such a decision will increase the likelihood that it will achieve its goals.

In these circumstances, this kind of decision-making strategy will always produce the best possible outcomes for the agent, no matter what particular outcome actually arises.

How can this kind of reasoning be applied to the foundations of science? As we shall see, it is applicable because science is faced with radical uncertainty. We will begin our analysis by considering an agent that fully accepts Hume's argument. As such, when the agent makes decisions about how it might act in order to advance its future interests, it knows that it faces radical uncertainty. It accepts that it is faced with an infinite number of possible outcomes in the next instant. It has no rational basis whatsoever to assign different possibilities to any of these possible outcomes. What outputs from science, if any, should the agent use in making its decisions? How should it act to achieve its goals optimally?

In answering these key questions, it is useful to begin by considering the possible outcomes for an agent if it makes its decisions on the basis of hypotheses that conflict with those established by science. In other words, it makes its decisions using the predictions of any of the infinite number of hypotheses that are inconsistent with the laws and other regularities that have been established by science.

The key point is this: Even if the predictions turn out to be correct, the agent may not benefit at all. This is because, if the universe suddenly begins to behave in a way that is inconsistent with the way it has in the past, life in the changed universe may no longer be possible. An agent who is extremely lucky to choose the correct one of an infinite number of possibilities may not survive the experience.

As I have mentioned, there is a lot of evidence that the laws and initial conditions of our universe appear to be fine-tuned in ways that enable life to arise and persist. If the regularities that underpin these laws were suddenly to cease to exist, the universe may cease to be life-friendly. Furthermore, evolution has taken advantage of these laws and regularities to build the processes that enable living processes to function effectively. Consequently, the continued functioning of our nervous, physiological, metabolic, cellular, and other processes depends on the continued existence of these regularities. If these past regularities were suddenly to cease to exist, so too would our lives.

Our science is not sufficiently advanced to identify in detail which of these past laws, regularities, and other patterns are essential to our continued existence. But the complex interrelationships of the constituents of our bodies and those between our bodies and our environment suggest that even slight changes to past patterns may be fatal, instantaneously.

Given these considerations, if a living agent bases its decisions on hypotheses that are inconsistent with past patterns, and if the predictions of these hypotheses suddenly emerge, the agent is unlikely to survive to appreciate any benefits from getting it right.

There is only one decision-making strategy that an agent can adopt that will ensure that it will benefit at least in some cases, no matter what hypotheses prove to be accurate. This strategy is to base its decisions on the meta-hypothesis that regularities that appear to have prevailed in the past will generally continue. Of course, a community of rational agents will hedge its bets somewhat by including research programs that test the boundaries of this hypothesis through time.

These considerations suggest that the rational reason why we should base our decisions broadly on the predictions of established science is not because they are more likely to be correct. Rather, it is because even if hypotheses that conflict with past patterns prove to make accurate predictions about the future, relying on these hypotheses will not get us anywhere. If they are correct, it is likely that we will not be around to enjoy any benefits that might flow from our accurate predictions.

This is not the place to develop in detail all the consequences and nuances of this proposed approach to the philosophy and epistemology of science. Suffice it to say that the arguments I have sketched justify scientific methodologies that are broadly similar to those practiced by the scientific community. There are important differences in detail, but it is beyond the scope of this book to consider these in greater depth.

Largely, science has been doing it right, but for entirely the wrong reasons.

These insights point the way toward the development of a rational approach to making decisions that applies equally to the methodology of science as well as to fundamental existential issues. However, the specific approach I have taken to making decisions that rely on scientific methodologies cannot be applied to decisions about events that occur after the agent has died. The methods that are specific to science cannot be applied in circumstances where the testing of hypotheses is not possible.

However, at a higher level of abstraction, the same kind of reasoning that was used in relation to science is equally applicable to decision-making about consequences that are not experienced until after the agent dies. Both cases deal with how an agent can make decisions rationally, even in the face of radical uncertainty about the future consequences of decisions.

Any conclusions reached by this reasoning in either domain will have equal validity. Conclusions reached about how an agent should act now in the face of post-death possibilities are no more nor less justified than science-based conclusions about how to act in relation to pre-death events. At a higher level of generality, this approach unifies science with existential reasoning. In neither case does it generate certain knowledge about future events. However, in both cases, it identifies how an intelligent and rational being should reason in order to undertake decision-making in the face of radical uncertainty.

HUMAN SUPERINTELLIGENCE

Part 3: How to Scaffold Metasystemic Cognition and Beyond

23.

How to Develop Skills that Enable Metasystemic Cognition

In the final two chapters of the book, I detail how an individual's capacity for metasystemic cognition can be scaffolded, including how to self-scaffold such an ability.

These chapters draw together the various threads introduced in earlier chapters about how higher cognition can be developed.

As I outlined in the Introduction, my strategy to this point in the book has been to embed information about cognitive development in a narrative about my own developmental odyssey. This strategy has multiple purposes.

First and foremost, the strategy was designed to overcome a fundamental difficulty confronting attempts to convey to others what they can do to achieve metasystemic cognition. The difficulty is that the practices and methods needed to develop metasystemic cognition can only be understood fully by individuals who are already equipped with it. This is because the processes involved in acquiring and implementing metasystemic cognition are highly complex. They can only be understood by an individual who can build mental models of highly dynamic, complex systems.

The Catch-22 is that in order to construct and operate these complex models, you need a capacity for metasystemic cognition. Understanding how metasystemic cognition can be developed requires building complex mental models of cognitive processes. But this obviously requires metasystemic cognition.

When individuals at the analytical/rational level are provided with insights that were generated by metasystemic thinking, they will often fail to understand the basis of those insights. This is because the metasystemic insights will have been derived from an understanding of factors and processes that are not included in the mental models of analytical/rational thinkers. The insights will not make sense. To analytical/rational thinkers, the conclusions reached by a metasystemic thinker will not seem to be justified. Broadly, this is how behaviour generated at a higher level of cognition will look to intelligence at a lower level.

This posed a fundamental challenge for this book: How can it overcome the Catch-22? How can it convey to readers an understanding of metasystemic cognition that is sufficient to enable them to scaffold it? In particular, how can it achieve this given that a full understanding requires metasystemic cognition, and very few readers will already have this capacity?

This challenge is compounded by the fact that much of the knowledge needed to develop a capacity for metasystemic cognition cannot be conveyed by declarative propositions and explanations alone. In part, this is because much of it is procedural knowledge. It is in the form of skills and practices that cannot be described or conveyed in detail in words.

Furthermore, there are no words available to describe many of the experiences and states that individuals need to recognise if they are to understand what is being conveyed. This limitation cannot be overcome easily. Creating a new word for a particular experience will not convey any meaning to individuals who have not had that experience. When the word is used, they will not understand from their own experience what it is referring to.

The strategy I adopted to answer this challenge, at least in part, was to provide a structured narrative about how one particular individual (myself) developed from analytical/rational thinking to metasystemic cognition over the course of his life. This narrative provides analytical/rational thinkers with an outline of the key steps that they need to take to develop metasystemic cognition.

Importantly, it provides this information in a graduated fashion. This enables new concepts and understandings to be introduced in as simple a form as possible, and then elaborated in greater depth and complexity as the narrative proceeds. The structured narrative will tend to mirror how individuals are likely to need to develop their understandings and concepts as they progress.

Overall, the narrative is structured as a spiral in which a particular issue is developed to a degree before the narrative moves on to other issues. Later in the narrative, the issue is then revisited in greater depth, and so on. Such a structure avoids providing all the information relevant to a particular issue at once, in a fully-developed form. The analytical/rational thinker would be unable to grasp such complex material until they have made some progress towards metasystemic cognition. Furthermore, this approach encourages the reader to think ahead of the text, and to begin to develop their own mental models of how their thinking and feeling might need to develop.

A further benefit of the narrative approach is that it enables some non-declarative forms of knowledge to be transmitted. To an extent, it evokes the capacity of the reader to identify with the protagonist and feel empathy for him. This enables the reader to feel their way into the experiences of the protagonist. Even when the reader finds the behaviour of the protagonist unattractive and uninspiring, this might prompt the reader to think and feel in depth about the relevant issues.

The use of metaphor and analogy further enhances this process in which mirror neurons are engaged and embodied experience is evoked vicariously. These devices may also help the reader to identify relevant states and feelings that they may have experienced in other contexts. These may be similar to the states that the narrator is attempting to point to. This can enable the reader to, for example, understand particular states even when they have not had the specific experiences referred to by the narrator.

These approaches can also help to engage emotions that provide the reader with the motivation to undertake the sustained work on themselves required to develop higher cognition.

Of course, a disadvantage of adopting such a structure for the book is that these final two chapters will unavoidably repeat some of the ideas that were first introduced in the narrative chapters of the book. But

I have minimized repetition as much as possible, and have written these final chapters on the assumption that readers have read and assimilated earlier chapters, at least to some extent. Overall, for readers who have the potential to develop higher cognition, I think the benefits of my approach will significantly outweigh the disadvantages.

For these reasons, I recommend that you do not skip the narrative chapters of the book and begin your reading here. If your cognitive 'center of gravity' is currently at the analytical/rational level, what follows is unlikely to make sense to you, unless you have first read and absorbed the earlier material.

I will begin by providing a broad overview of what you will need to do to scaffold metasystemic cognition. After this overview, I will consider in detail each of the specific steps that you will need to take.

Metasystemic cognition is the capacity to construct mental models of dynamic, complex phenomena. Examples include models of evolving economic systems, political systems, cognitive systems, other psychological systems, physiological systems, organisms, and ecosystems.

An individual who is equipped only with analytical/rational thinking will not be able to construct such models. Such an individual will be capable of constructing and utilizing mental models that can adequately represent only phenomena that are mechanistic and that can be analysed. Broadly, these are the phenomena that can be understood by modern science. Unfortunately, however, most of reality cannot be modelled effectively using such models.

As discussed earlier in the book, the development and spread of metasystemic cognition is essential if humanity is to survive the existential threats that we face currently. The drivers of environmental destruction, including global warming and the threat of nuclear war, are all dynamic, complex phenomena. They cannot be fully understood with analytical/rational thinking alone. Without metasystemic cognition, humanity will also be unable to identify and develop the complex interventions that are required in our political, economic, and environmental systems if we are to mitigate these existential threats.

More broadly, humanity must advance beyond analytical/rational thinking if we are to be capable of developing a comprehensive evolutionary worldview and using it to guide our future evolution. Such a worldview is essential if humanity and life on this planet are to survive and thrive indefinitely into the future.

As discussed earlier, developmental psychologist Robert Kegan has identified in broad terms what individuals must achieve if they are to advance from one developmental level to another. The critically important step is that processes that were part of the subject at one level must become object to a new subject at a higher level. This enables them to be manipulated and optimised consciously by the higher level.

We can apply this general insight to the development of metasystemic cognition from analytical/rational thinking: an analytical/rational thinker is psychologically embedded in their analytical/rational thinking. It is part of their subject. Processes that are part of an individual's subject are unconscious. The subject cannot see and modify them or watch them in real-time. They are not object to the subject.

As such, the analytical/rational thinker has no psychological distance from their thought processes. Consequently, they cannot see the limitations of their thinking. As a further consequence, they tend to believe that analytical/rational thinking can solve any problem that has the potential to be solved. Unable to perceive their thinking from the outside, they cannot see that analytical/rational thinking can only solve a very small proportion of the challenges we face in life, individually or collectively.

Nor can they consciously control their analytical/rational thinking or modify it. Their thinking has them, they do not have it. It is not a tool that they can freely choose to use or not to use, depending on the circumstances. Instead, they are slaves to analytical/rational thinking.

This all suggests that to develop higher-level cognition, individuals first need to dis-embed from analytical/rational thinking and constitute a new, higher-level subject. Their analytical/rational thinking will be object to this new subject.

Initially, this new, higher-level subject will largely be a 'silent witness' of analytical/rational thought processes. But in order for the individual to develop metasystemic cognition, this higher subject/self will then need to be equipped with new thought processes. Appropriate new thought processes will enable the higher subject/self to build mental models of complex phenomena.

As I have outlined earlier, the acquisition of these new thought processes can be assisted significantly by the thought forms that have been identified and developed by Otto Laske. Appropriate use of these thought forms can greatly facilitate the scaffolding of the knowledge and skills needed to equip the new subject/self with metasystemic cognition.

Of course, Laske describes these thought forms using declarative knowledge. Consequently, it will be necessary for the individual to convert them into procedural skills by practicing their use. This is largely because many aspects of complex phenomena that need to be represented in mental models cannot be represented declaratively. For example, they include processes and patterns for which words do not exist.

Eventually, with sufficient practice, you will be able to use the thought forms fluidly and largely unconsciously, like other procedural skills, once they are learned sufficiently. This will enable you to use the thought forms to construct and manipulate mental models of complex phenomena.

In the next chapter, I will deal in detail with practices that you can use to develop the procedural knowledge needed to apply the thought forms to build complex models and embody metasystemic cognition.

Developing a new, higher-level self that is dis-embedded from thought and feeling is a critically important step towards acquiring metasystemic cognition.

First, it enables analytical/rational thinking to be seen as object. This will enable the new self to see the limitations of this thinking, and to see what needs to be included in its cognitive processes in order to overcome those limitations.

Second, by dis-embedding from thought processes, the new subject can intervene in them and modify them. This will enable it to improve them in whatever ways are needed to build new thought processes that are better at constructing metasystemic models. Ultimately, this capacity will enable individuals to develop new and modified cognitive processes that overcome the limitations of their analytical/rational thinking.

Third, by dis-embedding from thought and feeling, the new self can access right-brain capacities such as pattern recognition and intuition. Access to these important enabling capacities would otherwise be blocked by absorption in thought and feeling. This is because embeddedness in thought and feeling exhausts the limited bandwidth of consciousness, preventing conscious access to other capacities. Access to these capacities is essential if the new self is to have the ability to include representations in its models of aspects of complex phenomena that are too fluid and ill-defined to be captured in words and propositions.

Fourth, the ability of the new self to dis-embed from emotions and feelings will enable individuals to free themselves from the dictates of their evolutionary past and conditioning. When they choose to do so, they will be able to move at right angles to their biological, cultural, and social predispositions and conditioning. This capacity is critically important for enabling individuals to implement the superior strategies identified by their higher cognition.

Whenever an individual's pre-existing desires, emotional predispositions, or motivations conflict with the best strategy, the individual will be able to escape those conflicts and act optimally. As has been discussed earlier, the full development and elaboration of such a capacity enables an individual to become what I have referred to as a self-evolving being.

This brings us to a key challenge for the development of metasystemic cognition: How can an individual actually dis-embed from analytical/rational cognition and from emotions and other feelings, and see them as object? How can an individual develop a new, higher-level subject?

Following this overview, I will deal in detail with practices and approaches that will enable you to build these capacities. However, it is important to emphasise that developing these abilities cannot be achieved using the intellect alone. By itself, the acquisition of declarative knowledge will not install the capacities in you. Their achievement necessitates the acquisition of experiential and procedural knowledge and skills. Like skills in general, these can be acquired only through appropriate practices and training.

In summary, two key capacities need to be scaffolded to enable metasystemic cognition. The first is to develop a new, higher-level subject to which thinking and feeling are object. The second is to install a capacity for metasystemic cognition in this new subject.

I will now deal in detail with how you can scaffold these capacities. The remainder of the book will outline the key practices and knowledge you need.

I will begin with the practices and knowledge required to dis-embed from one's existing thinking and feeling. The goal is to develop a new, higher-level subject/self that is not controlled by thinking and feeling, and that can therefore modify existing thought processes in order to correct their limitations and deficiencies. Building such a new self provides important enabling skills for the development of metasystemic cognition.

Up until recently, the main source of these kinds of practices has been the religious and spiritual traditions. Over thousands of years, the traditions have developed and spread various forms of meditation practice that can scaffold some of the key elements of the enabling capacities that we have identified.

The reason why the traditions developed these practices is not because they wanted to scaffold higher cognition in their followers. Rather, it is primarily because the practices are capable of producing altered states of consciousness. These altered states tend to be able to be used to reinforce the belief systems that are fundamental to the traditions.

The altered states are often experienced as 'profound' and as evidencing a reality different from what is experienced in ordinary life.

This provides fertile ground for generating and reinforcing beliefs involving gods, spirits, and other supernatural phenomena.

However, we now understand that these altered states are a product of the way in which human psychology is organised, rather than a reflection of a different reality. We can explain the states and their effects using a science-based understanding of human psychological functioning.

For example, most practices developed by the traditions produce altered states by training the capacity to dis-embed from thought and feeling. Prior to the development of such a capacity, individuals are embedded in their thoughts, emotions, and other feelings. Generally, as you can confirm in your own experience, when an individual is embedded in a sequence of thoughts, their awareness is fully absorbed in their thinking. There is no conscious awareness left over that enables them to simultaneously maintain awareness of things outside their thinking—e.g., awareness of their environment. Consequently, if you engage in serious thinking while you go for a walk in a park, the park disappears from your conscious awareness while you are thinking.

The reason why we get embedded in thoughts or feelings is that our consciousness has a very narrow bandwidth. As many experiments have demonstrated, conscious processing is serial and slow. So, when we are thinking, our train of thought completely fills the narrow bandwidth of consciousness. Consequently, we experience our awareness as contracting down and being filled by our thinking.

For similar reasons, embeddedness in thought blocks other information from consciousness, even though this information might be highly relevant to our thinking. For example, if we are heavily involved in thought, access to emotional information and other right-brain capacities tends to be blocked.

This experimentally-confirmed understanding[25] of the limited bandwidth of consciousness seems to conflict with the belief of the religious and spiritual traditions that consciousness is 'infinite' and 'unbounded'. However, this apparent disagreement exists only on the

---

[25] For details, see Stewart (2007)

surface. Experientially, we cannot detect the boundaries of our conscious awareness. We are unable to perceive a horizon at which consciousness ends. This is the sense in which it is 'infinite'. But these experiences do not contradict the fact that humans are able to give full conscious attention to only limited areas of their external or internal environment at any time.

Currently, human beings spend their lives embedded in their thinking and feeling. They live almost entirely in contracted awareness. Only rarely do circumstances arise that produce dis-embeddedness and enable them to experience uncontracted awareness.

When consciousness is uncontracted, very little of the narrow bandwidth of consciousness is taken up. Consciousness is not absorbed in thinking and feeling. Consciousness is now able to experience thoughts and feelings as objects that arise along with other objects in awareness. The individual is no longer embedded in their thoughts or feelings any more than they are usually embedded in perceptions of physical objects that arise in awareness.

These observations provide an experiential understanding of what happens when perceptions shift from being part of the subject to now becoming object. Currently, most individuals experience much of their external environment as object. Generally, their attention is not automatically absorbed in perceptions of physical objects. They are capable of giving attention to an external object, or freely taking attention away from it. Objects in their environment do not have them, they have the objects.

When individuals develop the capacity to see thoughts and feelings as object, they will experience them in much the same way as they do physical objects. Their attention will no longer be automatically absorbed by the thoughts and feelings. They will be capable of choosing to give attention to thinking or choosing to take attention away from it. Thought and feeling will no longer have them. Instead, thinking and feelings will become tools that they can choose to use or not, depending on the needs of the circumstances they encounter.

For these reasons, when an individual dis-embeds from thought and feeling, they will experience themselves as 'being in the present' and as 'abiding in spacious awareness'. They will experience their external environment as being more vivid. Dis-embedded individuals experience themselves and the world very differently from the way they did when they were 'normal'.

Importantly, this is an extremely pleasant and peaceful experience. Most of our disagreeable psychological experiences occur as a result of our thinking, negative emotions, or other feelings. Once we dis-embed from them, and accept them as objects that arise in awareness and then dissipate, life is largely enjoyable. As Shakespeare put it: "There is nothing either good or bad, but thinking makes it so."

Sustained dis-embedding from thinking, emotions, and other feelings will produce 'the peace that passes all understanding'.

As I have mentioned previously, the traditions often describe the state produced by dis-embedding as awakening. This is because it is experientially analogous to awakening from dreaming sleep. In the dream state, awareness is embedded in the dream and its 'logic'. There is no wider, surplus awareness that enables the individual to evaluate their behaviour in the dream in the light of wider perspectives from outside the dream. The dreamer does not have access to a broader context. Typically, the dreamer acts in ways that would be seen as absurd when seen from such a wider context. But the dreamer is embedded in and absorbed by the dream. The dreamer believes the dream is real and that their dream behaviour makes sense.

However, when dreamers awaken from a dream, they dis-embed from it, and can see their dream behaviour from a wider context. They can reflect on it, evaluate it, and take into account wider knowledge and perspectives. The awakened dreamer soon sees that their dream was not real. In several respects, it was absurd and did not make sense.

When typical individuals develop the capacity to dis-embed from thinking and feeling, they are now able to see their previously embedded behaviour from a wider perspective. They realize that many of their earlier actions fail to make sense. Often, their former behaviour seems appropriate only from the limited perspective of the embedded state. From the wider, dis-embedded perspective that they can now inhabit, their previous behaviour will often seem absurd. It will be as if they have awakened from a dream.

Of course, very few human beings ever awaken from the dream of normal existence to any extent. They spend their entire life absorbed in and being jerked around by thoughts, emotions, and other feelings. Consequently, they never experience the fact that their normal existence is as absurd as their dreaming state appears once they awaken from sleep. You can experience what it is like to awaken only by doing the necessary work to develop the ability to awaken and sustain the awakening.

The religious and spiritual traditions founded their worldviews on the existence of these impactful states. Typically, each tradition suggested to its adherents that the profound experiences associated with the states demonstrated the validity of the tradition's belief system. Each tradition claimed that the profundity of the experiences provided strong evidence that their particular supernatural beliefs were correct, whatever they were.

But in order to be taken seriously, the traditions had to go beyond merely talking about these states and experiences. They had to develop practices enabling their followers to experience the states. Not only did they have to tell possible followers about the significance of these states, they had to enable them to access them.

Appropriate rituals, praying, collective hymn singing, worshipping in grand cathedrals, drugs, and other methods could all, at least momentarily, quieten the thinking mind and provide a taste of the altered state that arises when an individual is not embedded in thinking and feeling. But by far, the most successful forms of psycho-technology that have been developed by the traditions to produce awakening are meditation-like practices.

Religious and spiritual traditions have had to develop and teach appropriate meditation-like practices in order to deliver on their promises to provide adherents with transcendent experiences during their lives.

Christianity appears to be an exception. However, during its early development, it relied heavily on experiential practices, and several of the statements attributed to Jesus Christ by the Bible can be readily interpreted as referring to the kind of states I am pointing to. For example: "Be still and know that I am God,"; "The kingdom of God is within you,"; "Unless you change and become like little children, you will never enter the kingdom of heaven,"; and so on.

But in its violent attempts to exterminate Gnosticism in the third century AD, Christianity gave itself a spiritual lobotomy. It destroyed the esoteric elements in Christianity that had focused on practices that produced awakening in this life, here on Earth. Consequently, without practices that could produce awakening, Christianity had no alternative but to focus on promising its followers access to transcendence after they died. So far, they have gotten away with it, to some extent.

At their core, the meditation-like practices developed by the traditions contain a common element. This core element can be combined with other processes to produce a set of practices that can potentiate the scaffolding of metasystemic cognition.

This core element requires the individual to practice disengaging repeatedly from thought and feeling. The practice begins with the individual resting attention on sensations that are 'inert' in the sense that they do not tend to generate thinking or feeling. The sensations of the breath are a common example. Resting conscious attention on inert sensations tends to fill the limited bandwidth of consciousness with bare sensations, thereby helping to exclude thoughts and feelings.

When individuals find that they have again become embedded in thought or feeling, they gently and non-judgementally disengage from it and move their attention back to the inert sensations. This process of disengagement is repeated whenever meditators find that they have again become embedded in thought or feeling.

Initially, this practice is made easier by undertaking it in an environment that is free from distractions. The meditator can also reduce the potential for distractions by relaxing and dropping all intentions at the beginning of the meditation and during it.

As the meditation proceeds, repetition of this process tends to still the meditator's mind, making it easier to remain dis-embedded and experience the altered states that this produces.

Regular daily practice strengthens the meditator's ability to dis-embed and remain dis-embedded.

The traditions developed very effective practices for producing the transcendent states associated with dis-embedding. However, they did not develop additional practices for producing higher cognition and enhanced agency. As I have previously discussed, the traditions pursued the right-hand path. Its goal is absorption in the absolute, and its maxim is "thy will be done". In general, they were not interested in producing self-mastery and enhancing agency in this world.

The traditions had no interest in scaffolding higher cognition in order to enable adherents to act more strategically and intelligently in the real world. Their belief systems were focused on transcendence. In particular, the form of awakening they pursued was passive and surrendered. It was not active. They were happy for the higher-level subject to remain a surrendered, silent witness, rather than to equip it with higher cognition.

As such, the meditation-like processes developed by the traditions need to be further modified and extended if they are to be capable of scaffolding self-mastery and higher cognition.

Before we explore how this can be done, it is worth mentioning that the explanations given by the traditions for why their practices work cannot be relied upon. Every tradition provides such explanations. But across traditions, the explanations are often inconsistent and mutually contradictory. Furthermore, the explanations often rely on supernatural phenomena. In addition, each tradition relies on the particular supernatural phenomena that are peculiar to its worldview, contributing further to the contradictions between traditions.

The core practice developed by many traditions is effective. But in order to understand why and how it functions, it is necessary to take a science-based approach, and to strip from the core practice the supernatural explanations given by the various traditions.

It is then necessary to extend the core practice in several ways if it is to be capable of scaffolding enhanced agency and higher cognitive capacities.

These extensions include modifying the practice so that it can be used 'off the meditation cushion', in the midst of ordinary life. First and foremost, this requires the practitioner to learn how to use a source of 'insert' sensations that are 'portable'—i.e., that can be used effectively while the practitioner is engaged in everyday experiences, not just when sitting on a meditation cushion in a quiet room.

Many find that resting attention on the sensations of the breath does not work very well for this purpose. This is because these sensations are changing continually, and thus can generate distractions. As I have mentioned earlier, I also found that resting dis-embedded attention on the external environment was ineffective for maintaining dis-embeddedness for extended periods—the external environment also provides many distractions. Collective practices undertaken in a quiet environment also have limitations—they do not translate easily to situations where you are on your own in the midst of ordinary life.

This inert source of sensations will need to be suitable for 'anchoring' the individual in the dis-embedded state. By this I mean that when the individual is in a dis-embedded state in the midst of ordinary life, they will need to be able to keep part of their attention resting on the source of inert sensations in their body. Doing so will help to ensure that this part of their attention remains shielded from potential distractions, unlike the remainder of their attention which is out in their external environment. Achieving this will tend to anchor the individual in the awakened state, and help to prevent them from re-embedding in thought and feeling.

A further way in which the core practice needs to be extended is to use it in the service of agentic goals. At first, this may seem paradoxical because part of the core practice is surrendering and letting go of goals and motivations. However, once this part of the practice has contributed sufficiently to achieving dis-embedding, the individual can reintroduce goals and agency, but not in a way that engenders re-embedding. The witness will continue to be silent, but it will now be goal-oriented.

This is a very important modification of the core practice, given that the overarching goal of the modified practice is to enhance agency and intelligence.

What kinds of practices can assist the development of such a capacity for dis-embedded agency?

Individuals can use dis-embedding consciously and intentionally to disengage from strong negative emotions that are maladaptive. They can train themselves to come into the present whenever they encounter emotionally-challenging circumstances and to watch without re-embedding as their emotions emerge in real-time and dissipate. This creates a 'window of opportunity' in which they can choose to act in more adaptive ways. It gives them the capacity to move at right angles to their emotional predispositions.

A similar practice can be used in circumstances that produce maladaptive thinking in the form of unproductive worry and catastrophizing.

Like all practices that are repeated often enough, these kinds of responses to challenges will eventually become automatic and will not require conscious attention. Whenever the individual encounters circumstances that previously triggered negative emotions or thinking, the circumstances will now evoke a learned response to dis-embed and remain dis-embedded.

In some cases, thoughts and emotions produced by particular challenges might be overwhelming. These challenges might contract awareness and defeat attempts to dis-embed and to remain present. In these cases, instead of trying to use the practice when the challenges actually arise in real life, individuals can undertake the practice in their imagination.

The individual can use visualization and related techniques to imagine encountering these kinds of challenges, and can undertake the practice in these imagined circumstances where the thoughts and emotions are not so overwhelming. A possibility that is becoming increasingly available is to begin by undertaking the practice in an appropriate virtual environment.

An important goal of these practices is to be able to become and remain dis-embedded in the face of higher levels of distraction. To develop this capacity, the individual can begin by practicing in a quiet environment, 'on the meditation cushion'. Then, they can set out intentionally to use the practice in circumstances that are progressively more distracting. The ultimate goal is to train the ability to become dis-embedded and to remain so in any circumstances, no matter how triggering or challenging those circumstances may have been previously.

The achievement of this goal has been referred to as 'subject permanence'. When this has been attained, the individual is continually awake. Gurdjieff describes it as the emergence of 'permanent I'. It is what I am referring to here as the emergence of a permanently dis-embedded, higher-level subject or self that can consciously and intentionally manage the pre-existing psychological processes of the individual into a coherent and agentic whole.

In his characteristically provocative fashion, Gurdjieff suggested that until individuals develop such an 'I' that continues to be dis-embedded in the face of distractions, the individual does not possess any 'I' that is psychologically enduring. Their 'I' winks in and out of existence as they dis-embed and then re-embed. They do not have anything permanent that has the potential to continue in existence indefinitely.

Gurdjieff went on to draw the conclusion that humans are not born with a soul. A soul is something that humans have to build and develop intentionally during their lives through intense work on themselves. At present, almost no humans do so.

It is also worth noting here that when this higher-level subject emerges, it tends to experience itself as standing outside time. This is because it is not embedded in the perceptions, thoughts, and emotions that deal with events in time and mark time's passage by changing continually. Instead, it stands outside and is unaffected by these events. The higher self experiences itself as awareness, the ground of all perceptions, and this awareness never changes. As the Bhagavad Gita states, it experiences itself as having never been born and never dying.

Consequently, the new subject can experience itself as existing outside time, in eternity. When resting in this new self, individuals can experience themselves as having eternal life. Arguably, this is the eternal life promised by the Christian Bible. Of course, it is not the same as actually achieving immortality.

Of course, these kinds of metaphorical descriptions will often be misinterpreted by those who have never had the experiences that these metaphors are pointing to. Taking the metaphors literally can never produce an accurate understanding of what they are attempting to describe. If an individual has not had the relevant experiences, no words exist that can produce those experiences in the individual. Words or symbols alone are incapable of enabling them to experience what is being pointed to.

Reading books about what sugar tastes like will not enable an individual to experience the taste of sugar. Nor will scientific explanations about how our taste buds function. The only way we can ever experience what sugar tastes like is to actually taste the sugar.

It is for these reasons that the original teachings of the founders of the great spiritual and religious traditions are often misunderstood by their followers in subsequent generations. Rarely do the founders themselves actually write anything about their teachings. What they offer is not words. Instead, they offer access to impactful experiences and altered states.

The founders of the traditions seem to understand that these experiences cannot be conveyed in words to those who have never had the experiences. Some actually predict that their teachings will be misunderstood, except by the tiny minority who work intensely on themselves with the relevant practices. Their teachings can only be understood properly by those who actually taste the sugar.

As I have mentioned, it is possible to extend meditation-like practices in ways that improve their potential to enhance agency. The extended practices that are most relevant for the purposes of this book are those that can be used to scaffold higher cognition. In these practices, the individual intentionally and consciously uses dis-embedding as a tool for building enhanced cognition.

An important example is the use of dis-embedding to access what I have referred to somewhat metaphorically as 'right-brain capacities'. We have discussed how these resources are essential for embodying metasystemic cognition. Foremost amongst these are capacities that enable us to recognise and use patterns and processes that cannot be represented adequately with words. In order to build mental models of complex phenomena, an individual must be able to represent processes and patterns in their model building.[26]

Complex dynamic processes and patterns are unanalysable, so they cannot be adequately represented in analytical/rational models. Analytical/rational thinking is highly reliant on declarative, left-brain resources.

In significant part, the development of metasystemic cognition necessitates the integration of right and left-brain capacities. The construction of mental models of complex, dynamic phenomena requires both capacities. Speaking somewhat metaphorically, metasystemic cognition requires the synthesis of right and left-brain processes, the integration of head and heart; the marriage of the bride and the groom; the integration of feeling and thinking; and so on.

This integration can be facilitated intentionally by making use of the fact that dis-embedding from thinking is capable of 'unblocking' access to right-brain resources. When individuals are embedded in their thoughts, the narrow bandwidth of consciousness is fully loaded with thinking. This blocks the conscious mind from recruiting relevant resources from the unconscious mind and right-brain processes. Dis-embedding unloads consciousness, and enables these resources to emerge into conscious awareness as, for example, intuition or insight.

The potential of circumstances that produce dis-embedding to enhance cognition functions such as problem-solving has long been recognised. However, their ability to do so has not generally been

---

[26] For comprehensive details about right and left-brain capacities, see McGilchrist (2009, 2021). However, McGilchrist's books do not contain detailed practices and other technologies that enable individuals to integrate these disparate capacities, including in the service of higher cognition.

explained in these terms. Taking a shower, walking in a natural environment, having a nap, or listening to certain kinds of music, can all quieten the mind and reduce the ability of thinking to dominate consciousness.

However, meditation-like dis-embedding practices are far more effective and convenient. They enable an individual to dis-embed at any time and in any circumstance. Most importantly, they enable the individual to intersperse thinking with unloaded, uncontracted awareness. Eventually, continued use of the practice will facilitate the fluid and effortless integration of conscious thinking with uncontracted awareness, enabling access to unconscious processing whenever it is useful.

A very effective method for enhancing the agentic and cognitive capacities of the new self is to use them to build mental models of complex phenomena. Building your own mental models and recursively improving them exercises the capacity to see your thought processes as object, to evaluate them, and to manipulate them consciously.

Individuals who want to enhance their capacity to dis-embed intentionally and to install their higher-level self with metasystemic cognition will actively look for opportunities to build mental models of complex phenomena from first principles and to improve them recursively. Such a practice is central to the development of metasystemic cognition. In the next chapter, I will deal in detail with these kinds of self-scaffolding practices.

The more skills are practiced, the more their use will become automatic, and the less that they will absorb consciousness and contract awareness. It is like learning to drive a car. Initially, it requires focused attention. However, individuals will eventually be able to drive the car largely automatically, with conscious attention directed elsewhere, unless circumstances arise that cannot be dealt with effectively by their learned, automatic behaviours.

As I have outlined, the original core practices developed by the religious and spiritual traditions can be enhanced and modified to produce new forms of practice that are purpose-built to enhance agency and cognitive capacity. What would these new practices look like in detail?

I will now set out to provide a general guide to the central elements of extended dis-embedding practices.

Of course, this guide should be used in the same way that a manual that attempts to teach a better tennis serve should be used. Reading and fully understanding everything in the manual will not by itself enable you to perform the enhanced service action described in the manual. Nor will achieving one hundred percent on an examination about the manual's contents.

Instead, in order to use the tennis-serving manual effectively, you will need to try out the particular movements that the manual is attempting to describe. This will involve experimentation and trial-and-error. This is because the words in the manual cannot specify precisely how you should perform each movement. This problem is exacerbated by the fact that everyone is different, so the precise movements that are optimal for one individual are unlikely to be optimal for another. The manual cannot specify particular movements that will work for everyone.

Initially, your attempts will probably not get you far. They could make your overall serve worse. You will have begun by experimenting with different movements, but need some way to evaluate whether you are headed in the right direction. The key step that you will need to take to get past this point is to have some way to judge whether the new movements that you are trying out are, in fact, the 'correct' movements.

If you are to make progress, you will need some method of assessing whether your attempts to implement the required movements are successful or not. You will have begun by experimenting with alternatives using trial-and-error, but then you will need to be guided by a feedback mechanism. This feedback will need to inform you when trial-and-error has produced the actual kind of movement that is intended by the manual, and that works for you.

An effective manual will help you to develop and utilize such a feedback mechanism. The manual will assist you to recognise a feeling in your body or some 'shape' to your movements that will reliably indicate that you are perfecting the movement. This will enable you to keep trying out different approaches until you achieve the outcome identified by the manual. The manual will provide you with practical indicators that enable you to judge whether you are performing a movement in the way that it is attempting to teach. You will learn from the consequences of your trial-and-error explorations.

This process will be more effective if the manual provides an indicator of success for each specific movement involved in producing an effective serve. This will be far more useful than if the only indicator it provides is for the serve in its entirety.

Ideally, indicators should be provided for each key movement in the serve, then for each relevant combination of these movements, and so on, producing a hierarchy of indicators.

In summary, when attempting to use the practices I will now outline, remember that you will not 'get them right' easily. You will need to be extremely persistent, explore with trial-and-error, and continually evaluate your attempts by testing them against their intended effects. If you find that the practices that I outline do not seem to be working for you, the answer will rarely be found in the guide I am outlining. Rather, it will be found by experimenting and closely watching the consequences.

There are many, many particular ways to undertake the core disengaging practice that the religious and spiritual traditions have developed. Thousands of guided meditations are available on YouTube. These can be useful to begin with. However, the key goal is to transition as quickly as possible to put yourself in the driver's seat in relation to your practices. This does not conflict with the need to surrender and drop all intentions at particular points in your practices. Surrendering can be a means for achieving an end that is being pursued intentionally.

Most examples of the core practice begin with the practitioner sitting comfortably and upright on a chair in a quiet room with minimal distractions. Environments that produce external distractions are avoided because they tend to produce re-embeddedness in thoughts and feelings.

But keep in mind that the ultimate goal of the practitioner who is attempting to enhance agency is to develop the capacity to dis-embed and remain dis-embedded in a highly distracting environment, including in the midst of all the kinds of distractions that arise in ordinary life.

Achieving dis-embeddedness can be further facilitated by preliminary practices and postures that quieten the mind. This reduces internal distractions. These preliminary practices can include, for example, giving attention to a variety of inert sensations for a short period.

This can include moving attention around a number of sources of sensation and pausing to rest attention on each in turn. For example, attention can be given to the feeling of air on your skin, the sensations where your body touches the surfaces that you are sitting on, the sensations arising from the soles of your feet on the floor, noises arising in your environment, and so on. As you rest attention on such sensations, you can relax further, noticing that this often makes sensations easier to detect.

If you find yourself re-embedded in thought or feeling at any time, you gently and non-judgementally move your attention back to the sensations. As we have discussed, repeated disengagement and return to inert sensations are central to the dis-embedding practice.

It is important to accept that continually finding yourself re-embedded in thought and feeling is not a failure of your meditation. In fact, it indicates that you are not wasting your time. The purpose of the practice is to strengthen your ability to disengage from thought and feeling and to remain disengaged. This strengthening occurs by repeatedly practicing dis-embedding.

The reason why you are doing the practice is to disengage repeatedly. If you do not find that you are re-embedding regularly and that you need to disengage, you are not practicing. You are not building up your capacity to disengage and to remain disengaged.

If you find that you have begun to think about these issues during the practice and are mulling over whether you are doing it right or wasting your time, you have re-embedded in thought or feeling. Whenever you notice this occurring, you use the practice: you disengage your attention from the thinking and move back to sensations, gently and non-judgementally.

The test of whether you have been doing these preliminary practices effectively is whether your mind is quieting to some extent, and you are relaxing and surrendering.

After these preliminary preparations have achieved this outcome, the next step is to begin the main part of the core practice. This involves concentrating attention for a prolonged period on a source of inert sensations within your body.

Perhaps the source most used by the traditions is the sensations associated with the breath. But in principle, any source of inert sensations will suffice that persist over the course of the meditation practice. These include sensations within parts of the body, including, for example, in the feet, arms, or fingers. Or sensations in the head. This works whether the sensations are real or imagined or a combination.

However, given that your ultimate goal is to use the practice to dis-embed and awaken in the midst of ordinary life, you need a source of sensations that you can use to anchor yourself successfully as you function in daily activities. In particular, you need a source of internal sensations that can be the focus of part of your attention while you are out and about in the world. As I have mentioned, giving attention to sensations associated with the breath does not achieve this effectively for me. The continual changes in these sensations as I breathe in and out and as breathing changes in different circumstances give rise to distractions.

The source of sensations that work well for me on both the meditation cushion and in daily life are sensations in my forehead, just above or below the level of my eyes.

Initially during the practice, I focus my attention on these sensations. This is a form of concentration meditation, that contrasts with a mindfulness approach. Mindfulness is generally associated with attention that is less concentrated and more open-focused. As a person who tends to live in my head, thinking continually, I found that the concentration approach was more effective. It could quickly and powerfully quieten my mind to a level where the disengaging practice was easier to perform.

To recap, the core of the practice is to concentrate attention on the source of sensations; to further relax the body and to sink into the feelings of relaxation; to disengage attention whenever you find that you are re-embedded; to then move attention back to the source of sensations; and to do so repeatedly. You should apply the same processes if you find yourself embedded in meta-thinking about whether the practice is working, and so on.

When I started this practice over thirty years ago, it would progressively quieten my mind. Then, after about twenty minutes, I would experience a significant state change in my awareness.

Before I had started the practice, I had come across a simple attempt to describe the experience that arises when one undergoes this kind of state transition. The description tallied with my experience of the transition, and reassured me that I was on the right track. It gave me a useful indicator of when my practice was 'working'. The observation that provided this indicator was that when the transition occurs, the experience is always surprising and unmistakable.

It was, and it still is after all this time. It is as if all disturbances within awareness that are typically caused by thought or feeling are quelled. Conscious awareness becomes a solid block. As I have mentioned, a friend describes the transition aptly as "the lake freezing over". The transition is accompanied by calmness and peace. As I have recounted earlier, I now finally know what the Bible points to when it refers to "the peace that passes all understanding."

Because the state tends not to contain significant internal disturbances, it is relatively easy to maintain without re-embedding in thought and feeling. But it is far from permanent. Re-embedding will still occur regularly, particularly as you move out into the world.

Continued daily practice (preferably morning and evening) will shorten the time that it takes for this transition to occur within any particular meditation session.

Given that your goal is to transfer these capacities to ordinary life, it is useful at the end of each meditation session to experiment with attempting to maintain the state when you open your eyes. It is also useful to experiment with performing the practice from the outset with eyes wide open. You will find that it is still possible to concentrate part of your attention on sensations in your head, even if you have your eyes open.

Furthermore, once the lake freezes over after a period of concentration meditation, it is possible to begin to practice expanding your awareness while you continue with the meditation. When you start this additional practice, your 'frozen' attention is concentrated on a localized area in your head. Then, you can practice expanding your attention outwards to encompass the inside of your head as a whole. As your practice develops, it will prove useful to experience this as giving attention to attention and to develop the ability to access the experience at will. To achieve this, you will need to practice resting indefinitely in this experience of giving attention to the source of your attention.

As a further preparation for awakening in ordinary life, you can then practice expanding this attention to encompass the remainder of your body. It is worth emphasising here that when attention is expanded, its focus is not shifted from one location to another. Instead, it is expanded from where it was originally localized in order to encompass the new location(s) as well. Because it is expanded, not shifted, it retains awareness of its original location.

The final transition that you will need to practice is to expand your awareness further so that it now encompasses your environment in addition to your head and body. In this state of uncontracted, dis-embedded awareness, you will be able to see your internal thoughts and feelings as well as your external environment as object. It is extremely useful for enabling self-observation of yourself and your reactions to events, as well as for monitoring and evaluating your thought processes.

Once you have practiced and entrenched the ability to do this, you can then begin to use it to further extend the awakened state beyond the end of the formal, sitting meditation. As a first step, when you open your eyes, you can practice continuing to be aware of your head and body, but not of the environment. You will find this easier to maintain than expanding your awareness immediately to include your environment. The environment will tend to be a source of distractions.

As you apply and experiment with these practices, you will increasingly learn to be in the world with your eyes open in spacious awareness that accepts and surrenders to whatever arises in your environment. However, retaining part of your attention anchored in your body on your attention is critically important for maintaining such a state.

As you practice your way through these phases, you will likely encounter several difficulties and challenges. As very much a head-orientated person, I found it relatively easy to rest attention on attention in my head, and then to expand this into spacious awareness of my head and my environment. I also had little difficulty expanding awareness to encompass the space within my body. But this awareness of my body tended to be empty of feelings and emotions. Even in the midst of ordinary life, internal bodily feelings and sensations did not figure much in my awareness.

However, my bout of general anxiety disorder and the anxiety-like symptoms that I experienced when withdrawing from mood stabilizers helped change this. It propelled me to give far greater attention to what was happening in my body than I had ever done before. The need to deal with these extremely negative experiences drove me to develop the capacity to give full attention to my emotional system and other internal bodily processes, and to radically accept and surrender to whatever arose.

The resources that helped me most to achieve these goals were Judith Blackstone's book 'Trauma and the Unbound Body'[27] and her guided meditations, which are available on YouTube. Broadly, her practices involve using surrendered attention to identify and inhabit any areas of the body in which relevant emotions and related feelings cannot be observed, experienced, and radically accepted. The practices help to surface these blockages and deal with them, eventually enabling expanded awareness to fully inhabit the body and the processes that arise within it.

---

[27] Blackstone (2018) – see References

As Judith Blackstone points out, we might learn to cope with childhood trauma by suppressing awareness of the painful emotions that the trauma triggers within our bodies. One method that we might use to achieve this is to fill the limited bandwidth of consciousness with incessant thinking. Fully embedded in thought, we no longer feel the painful emotions. Problem solved! This would work fine if emotions did not provide us with an important source of information that helps us to adapt effectively, particularly in social situations.

Learning to remove these kinds of blockages is very important for enhancing agency in the world. As a teenager, I was intrigued by the Gurdjieffian idea that a key goal in working on oneself should be to have all the psychological centers that comprise a human being working together. The three main centers he refers to are the intellectual center, which is experienced predominantly in the head; the emotional center, which is experienced predominantly in the heart; and the physical center, which is experienced primarily in the abdomen.

Gurdjieff argued that it is only when these three centers work together that individuals maximize their potential to adapt effectively to challenges. Only then can they access all the relevant resources of the three centers. If their response is to be optimal, it needs to be able to call on whatever combination of head, heart, and physical resources is optimal.

However, Gurdjieff's writing did not outline in detail any practices that could be used to develop this capacity.

I realized after I began to use Judith Blackstone's practices that they provide a royal road to achieving the mode of functioning that Gurdjieff was pointing to. Before long, in the midst of ordinary life, I could rest in spacious, dis-embedded awareness that encompassed my head and body. In this state, I could experience the openness and sensitivity of each of the three centers—intellectual, emotional, and physical. In particular, as Judith Blackstone promises, by inhabiting my heart and chest, I could now experience deeper, more fluid emotional responsiveness to my environment, including to people I interacted with.

Previously, I could enter and rest in spacious, dis-embedded awareness at will. But to a large extent, this awareness encompassed

348

only my mind and environment. Once I had dis-embedded from thinking, I was in spacious, peaceful awareness. I interpreted this experience as being fully in the present and awakened. But it was not until I had to deal with anxiety and came across Judith Blackstone's practices that I discovered that much more was required. In addition to being present and dis-embedded, it is also necessary to have full access to all the resources of intellect, emotion, and the body. Awareness must encompass all three centers, simultaneously.

The development of this capacity significantly improved my agency and effectiveness in the world. In this state, I could draw on whatever intellectual, emotional, and physical resources I possessed that were relevant to the challenges I encountered. I was not just present passively. Instead, I could now access a state of dis-embedded presence in which I was silently and peacefully poised to respond optimally to whatever challenges arose.

Many individuals have internal blockages that impede the free expansion and movement of awareness around their body and mind. However, childhood traumas or other causes of blockages are likely to vary among individuals. Furthermore, the individual is probably unaware of the specific parental behaviour or other events that produced them.

However, as Blackstone suggests, if you give the blockages dis-embedded attention, they and their underlying causes will tend to surface into conscious awareness. Then, when the effects of past trauma surface, you can continue to accept whatever arises and witness it fully in dis-embedded awareness. This will tend to extinguish any maladaptive learning, conditioning, and suppression that was produced by the trauma.

This process can be accelerated by using imagination and visualization to intensify the re-experiencing. Used in conjunction with dis-embedded awareness, this enables trauma to be re-experienced and discharged in relative safety and calm.

Once these kinds of blockages have been removed, the next step is to practice dis-embedding and coming into the present in the midst of daily life, and maintaining that awakening.

There are several approaches that you can experiment with and practice. One is to begin by focusing at least part of your attention on the sources of inert sensations that you use during meditation. This is followed by expanding your awareness to also encompass the body, sensing the three centers. Finally, you expand awareness further to include the environment as a whole (it is worth emphasising here that this expanded awareness is not focused on particular objects or events in the environment. Instead, awareness is spacious and dis-embedded). If your meditation and other practices that train the elements of this approach have been effective, you will be able to do this quickly and safely, even while operating heavy machinery.

Another approach is to begin by dis-embedding attention from specific events, and then expanding it to encompass your environment as a whole, as well as your body. When I do this, my awareness tends to be associated with the space in which objects and events arise, not with objects and events themselves. As such, it tends not to be distracted by events.

A practice that can potentiate this experience and also anchor you in an awakened state is to continually maintain some attention on the space behind and above yourself. This can be done while walking around in a park or on city streets. Such a practice will tend to reduce distractions, and enable you to abide in spacious, dis-embedded awareness.

In this awakened state, the world will tend to look vibrant and almost alive. As I have mentioned, it is what Federico Fellini was referring to when he said that if you see with innocent eyes, everything is divine. 'Innocent eyes' refers to the fact that in this awakened state, the conscious mind is not thinking about events or objects, nor interpreting or judging them.

This is a very pleasant and enjoyable state, but it is not your end goal if you want to work on yourself in order to enhance agency and cognition. You can enter this peaceful state and maintain it simply by surrendering intentionality and accepting whatever arises. But if you want to retain and enhance agency, you will need to maintain intentionality even though you continue to be dis-embedded and in presence.

Gurdjieff's state of 'self-remembering' becomes highly relevant here. This is a state of awakening in which you experience yourself as an agent who is poised and ready to act in the world as necessary to achieve your goals and intentions. In this state, you are not surrendering agency and passively accepting whatever arises. Intentionality has re-entered the picture.

It is important to avoid the trap of valuing presence and awakening simply because it produces happiness and ease. Instead, you should value it because it constitutes a poised state in which you have access to all intellectual, emotional, and physical resources that might be useful for achieving your goals in whatever circumstances you encounter.

The first step towards experiencing self-remembering is to dis-embed and to focus your awareness on the source of awareness in your head. Then, you expand this awareness to include the space within your entire body, including all three centers—intellectual, emotional, and physical. This will enable you to sense each of these in the space of awareness within your body. Then, you expand your awareness to also include the space that encompasses your external environment.

The most important step follows this: you focus part of your dis-embedded attention specifically on your body, and simultaneously experience yourself as an embodied being that exists within a spacious environment. Done correctly, this will produce a direct and immediate experiential realization that you are an intentional being within an expansive environment, and that you are poised to act as you choose. This experience can be accentuated by simultaneously feeling in your three centers that "I am".

You can further potentiate this experience by breathing deeply into the awareness of your body at the same time that you move focused attention to it. This will heighten your awareness of your body and your mind, and further enliven the experience of "I am". The image that comes to my mind when I attempt to describe this is when we blow on dying embers in a fire in order to re-kindle them—our breath causes the embers to light up and come alive.

Performed effectively, this practice will produce a sudden and profound experience of 'being-in-the-world'. Suddenly, you will feel that you exist, that you have agency and intentions, that you are capable and effective, and that you are poised in stillness, ready to do whatever you need to in the world.

This realization contrasts with the experience of normal human beings who are continually lost in embeddedness in one event after another. When embedded, you forget yourself and your larger intentions. Hence Gurdjieff's term 'self-remembering'.

You will probably need to play around with these practices and experiment with them before you experience self-remembering. Initially, you will probably find it useful to practice resting attention on just the source of attention in your head, or on your emotional center, or on your physical center, or on your spacious environment as a whole, or on various combinations of these. Once you master these elements, you can attempt to put them together in the way I have described.

But you will know when you succeed in experiencing self-remembering. It is a powerful and unmistakable experience. It seems to be similar to the experience that a toddler has when it stands in front of a mirror and suddenly realizes that it is looking at itself.

With practice, you will be able to enter this state at any time during the day, in whatever circumstances arise. It is the most resourceful state you can enter for meeting whatever challenges you encounter. While you remember yourself, you will not get lost in external events or internal thoughts or feelings. Instead, because part of your attention will rest continually on yourself, you will be anchored in awareness of your own being.

In order to escape re-embedding, it is useful to maintain awareness of your goals and intentions. You need to be constantly aware that you are an agent in the world with your own agenda and strategies. This will enable you to continually experience a more or less permanent 'I' or self that exists and is running the show.

It is worth emphasising here that the state of self-remembering is not the same as the non-dual state (also referred to as the enlightened state or unity consciousness or the oneness state, and so on). It is not a state that is pursued by, understood, or revered by the religious and spiritual traditions and by the right-hand path. It is not the state that I fell into in 2008 and soon abandoned.

In the non-dual state, there is no experience of being a self that is separate from an external world. All is experienced as one. The non-dual state is not a state that enhances agency or cognition.

In fact, in its most highly developed and permanent condition, it manifests as dysfunctionality in the world. A striking example is the Indian non-dual sage, Ramana Maharshi. He was so surrendered and absorbed that when he was sitting in his traditional, cross-legged pose, he would defecate on his legs whenever his body was moved to do so. Fortunately, he had disciples who would clean him up. But this did not concern him—in the non-dual state, everything seems perfect just the way it is, and Ramana abided in surrendered bliss.

Developing a capacity to remember yourself in the midst of ordinary life will turbocharge the efficacy of all the practices I have outlined that use dis-embedding to enhance agency and cognition.

First and foremost, you will be able to use self-remembering to enhance the effectiveness of practices that help scaffold a capacity to be self-evolving. As we have seen, the goal of these practices is to free yourself from the dictates of your evolutionary past and conditioning.

The ability to self-remember can also be used to enhance dis-embedding practices that, for example, enable you to stay present in meetings and during other interactions with people, including when playing sports (it will enable you to be much more aware of what is going on and how others are reacting in real-time); to access pattern recognition, intuitional, and other resources to enhance higher cognition; to see your thought processes as object, to identify where they are limited and to use your emerging higher self to correct these deficiencies; to enhance the ability of the new self to construct mental models of complex phenomena and to use the models to identify strategies that will enable you to achieve your goals; and to develop whatever additional skills are necessary to equip your new higher self with a comprehensive capacity for metasystemic cognition.

Ultimately, if you learn how to integrate self-remembering with metasystemic cognition and the evolutionary worldview, you will be able to experience 'evolutionary self-remembering'. This entails experiencing yourself as an evolutionary agent embedded in an evolutionary environment in which you are poised to intervene. Your perceptions of this environment will be shaped by your mental models of the large-scale evolutionary processes that operate on Earth and in the cosmos more widely. In a state of evolutionary self-remembering, you can scan this spacious evolutionary environment and identify how you might act to support the successful evolution of life in the universe.

However, developing and entrenching a capacity for self-remembering is not easy. It will require intense and prolonged work on yourself. As I have mentioned, Gurdjieff said that you will have to work on yourself so hard that the soles of your feet will sweat. This was an understatement. I have spent a lifetime trying to understand and develop these capacities and am still very much a work in progress.

However, as humanity begins to transition to metasystemic cognition, developing these capacities will require less effort. As we enter a Second Enlightenment, new practices and other supports will emerge to scaffold these higher abilities.

In the meantime, it will continue to be extremely challenging.

In the next chapter, I will outline in detail practices and approaches that can be used to scaffold metasystemic cognition. However, the fact that I have dealt first with dis-embedding capacities is not meant to imply that they need to be acquired before attempts are made to develop metasystemic cognition. If this were the case, very few would ever get to work on developing metasystemic cognition, and most of these would reach old age before they began.

The practices and efforts that are needed to develop both dis-embedding capacities and metasystemic cognition are often mutually reinforcing and synergistic. Consequently, it is more productive to work on the development of both capacities together.

24.

Scaffolding Metasystemic Cognition and Beyond

What practices and knowledge can scaffold metasystemic cognition in ourselves and others? In the previous chapter, I identified how to scaffold skills that help enable the development of metasystemic cognition. These included the ability to dis-embed from existing thought processes, enabling you to see them as object. This, in turn, enables you to evaluate your thinking, identify any deficiencies and limitations, and intervene in your thinking to overcome any limitations.

Ultimately, the goal is to develop a new higher-level self that stands outside and manages your thinking and emotions, and that can develop metasystemic cognition.

But as I emphasised at the end of the previous chapter, it is not necessary to develop the enabling capacities in full before you begin to work in earnest on acquiring metasystemic cognition. In fact, working on both at the same time can be optimal. This is because the learning processes can reinforce each other synergistically.

As you will discover, practicing the building of metasystemic models requires you to attempt to see your thinking as object. Doing so will also exercise and strengthen your ability to build metasystemic models. Furthermore, progress in developing either set of skills is likely to increase your motivation to work on the other—it will enable you to experience more clearly the benefits of developing the other skill set.

How can metasystemic cognition itself be scaffolded? First, we need to have a clear understanding of what metasystemic cognition entails. Broadly, it is the ability to build mental models of complex phenomena. Often, this will entail putting together mental representations of complex systems as they interact, develop, and evolve over time.

Once we build these mental models, we can use them to simulate possible interventions, and to use the simulations to predict their consequences. This enables us to develop interventions and strategies that can be used to achieve particular goals.

For example, the systems and processes that are currently producing environmental destruction on a global level are too complex to be modelled by analytical/rational thinking. The same applies to the complex interventions that are needed to modify these systems in order to stop their destructive impacts. However, metasystemic mental models of the relevant dynamical systems will enable alternative strategies and interventions to be simulated and tested, facilitating the identification of the most effective interventions.

The first major step in scaffolding metasystemic cognition is to see the limitations of your current level of cognition.

Until you take this step, you are likely to continue to believe that your current thinking is potentially capable of solving all problems that are capable of being solved by thought. You are unlikely to see that your thinking fails to take into account phenomena that are nonetheless highly relevant to many of the problems that are important to you.

You will fail to see what is absent from your thinking because it is absent from your thinking. You cannot see and consider what is missing from your thought processes.

Realizing that your current thinking has significant limitations may seriously disturb you. Until now, you have used your analytical/rational cognition to navigate your way through the world. You have generally assumed that your thinking has served you well. Now you are beginning to discover that it is seriously deficient and not fit for purpose.

However, without this kind of deep and continuing shock to your psychology, you are highly unlikely to be motivated sufficiently to make the substantial efforts that are essential if you are to develop a higher level of cognition. For this, you have to lose faith in your existing thinking.

In general, vertical development occurs only when the individual begins to encounter contradictions between their current beliefs and reality. Often this involves existential crises and breakdowns that occur due to the inability of the individual's existing cognitive and social/emotional capacities to cope with challenges they encounter.

How can the ability to see the limitations of analytical/rational thinking be scaffolded? We will begin by identifying how analytical/rational thinking functions. Our focus will be on what it can do effectively and what it tends to leave out in its attempts to model reality.

This understanding of analytical/rational thinking will help you see your thinking as object. These pointers will assist you to see the functioning of your own analytical/rational thinking and to see its limitations.

Ultimately, this will enable you to take the second major step in the development of metasystemic cognition. This second step is to identify what new representations you need to include in your thinking if you are to succeed in building mental models of complex phenomena. Once you see what your current mental models are leaving out, you will be able to move your attention to these absences and begin to represent them in your mental models.

As we have seen, analytical/rational thinking is the level of cognition that powered the emergence of science, industrialization, and technological development. As such, it works very effectively for developing the kinds of mechanistic models that underpin First Enlightenment science and for designing and building machine-based technologies.

However, as discussed earlier, analytical/rational thinking has failed miserably in developing the sciences of complex dynamic phenomena. In particular, mainstream science heavily critiques the efforts of the humanities to develop rigorous models of complex social systems. However, mainstream science itself has failed to develop models and approaches that are able to overcome the deficiencies that its critiques have identified.

These difficulties all have their origins in the limitations of analytical/rational thinking. Broadly, analytical/rational thinking is restricted to modelling only those parts of reality that are analysable.

In order for a phenomenon to be analysable, it must meet a number of criteria: the phenomenon must be able to be broken down into identifiable components; these components must be able to be precisely defined; they must interact together according to clearly defined laws or rules; the components must generally be unchanged by the interactions, or where they are changed, this occurs in ways that are clearly definable; the components can be concrete aspects of reality, or in more advanced forms of analytical/rational thinking, they can be abstractions; and the rules that specify the outcome of causal interactions may also be abstract rather than concrete.

Analytical/rational thinkers are able to construct models that adequately represent these mechanistic, analysable parts of reality. They can use these models to predict how these parts of reality will unfold through time, and how they will react to possible interventions. The models are fundamentally reductionist and mechanistic. Their consequences can be predicted by deducing the outcomes of the clearly-defined rules that govern interactions. Provided there are not too many components to keep track of, the consequences of the models can be 'thought through' mentally, step by step by step.

Of course, unless the characteristics of the components are defined clearly and unambiguously, such models cannot be thought through and their consequences cannot be deduced. The same applies to the rules and laws that govern the interactions between components. This is why analytical/rational thinkers typically want to start any debate by first defining all the relevant components of the applicable mental models in detail. We will see that this insistence tends to 'freeze' or to 'reify' the components of the models, and thereby limits their ability to adequately model parts of reality that are constituted by complex phenomena that are ceaselessly changing.

Analytical/rational models work extremely well for understanding and manipulating parts of reality that can be approximated by models that are mechanistic and analysable. Science has progressed successfully since the First Enlightenment by searching for and analysing those limited parts of reality that can be approximated by such models.

Whenever intelligent organisms emerge on a planet, we can expect that their science will initially make discoveries and build models that can be successfully applied to analysable parts of reality. However, the organisms' cognitive capacity will have to develop to the metasystemic level before they can develop a science of their social and political systems and other complex phenomena.

As I have mentioned previously, a further understanding of the inability of analytical/rational thinking to build models of complex reality can be had by comparing the world built by analytical/rational thinking with the 'natural' world. Currently, we are surrounded by objects that have been designed and built by analytical/rational thinking. In the main, these objects are utterly different from the objects and systems we find in the natural world.

For example, we live and work in buildings with box-like structures. The rooms in these buildings are characterised by flat surfaces that meet at right angles. We drive along streets that travel in straight or curved lines. In our daily life, we use various kinds of machines to achieve our goals. These typically include motor vehicles, dishwashers, vacuum cleaners, personal computers, and so on.

These are the kinds of things that analytical/rational thinking can conceive of and construct. Our machines and our built environment are generally built from simpler components that interact in pre-determined ways. The components are largely unchanged by these interactions, at least until they wear out or break. As such, they are analysable by analytical/rational thinking, and their design and functioning can be thought through.

This is the built environment you see when you walk around the center of a large city. Compare it with what you see when you walk through a natural ecosystem, such as a rainforest. You are not surrounded by box-like structures characterised by flat surfaces and right angles. Analysable mechanisms and machines are largely non-existent.

Instead, you find yourself embedded in what analytical/rational thinking tends to perceive as an unintelligible and buzzing confusion, characterised at all levels by ceaseless change. Organisms are not mechanistic or analysable, nor are their complex interactions and the systems they constitute.

We can compare these two contrasting environments with a third example that lies somewhere in between. A farming community still contains many organic systems and processes that are not machine-like. However, the use of analytical/rational thinking to design and construct the farms is evidenced by a significant reduction in the number of species and the division of the land into a limited variety of plots and fields.

In order to construct an environment that can be managed for human purposes, analytical/rational thinking has to design a farm as a mechanism that is analysable and that can be thought through. Because it is unable to form mental models of complex, dynamic phenomena, it has to 'tame' the complex natural environment, reducing it to something that analytical/rational thinking can understand and manage.

As I have mentioned earlier, attempts by mainstream science to develop a science of complexity have produced outcomes that are similar to those produced by attempts to farm ecosystems— analytical/rational thinking has had to reduce and 'assume away' much of the complexity. The models produced by these efforts are incapable of adequately reflecting the characteristics of complex dynamical systems. They are largely analytical/rational reductions of complex phenomena. Consequently, so-called complexity science is able to understand complex phenomena only in the few cases in which the behaviour of the phenomena can be approximated by mechanistic, analysable models.

But if we are to scaffold the transition from analytical/rational thinking to metasystemic cognition and beyond, we need to know more than the limitations of analytical/rational thinking. We need to understand what it is that analytical/rational thinking leaves out when it attempts to model complex phenomena, and how these absences can be remediated.

Once we know this, we will understand what analytical/rational thinkers must give attention to if they are to overcome the deficiencies of their previous thinking. This knowledge will enable analytical/rational thinkers to identify those additional aspects of complex phenomena that must be represented in their models. It will enable them to break out of the mechanistic cage of analytical/rational thinking.

As well as identifying the additional aspects of reality that they need to capture in their models, aspiring metasystemic thinkers will also need to know how to develop mental representations of these aspects. Not only do they need to know what is missing, but they also need to know how to incorporate it appropriately into their thinking and modelling.

Otto Laske's Thought Form Framework is invaluable for assisting the aspiring metasystemic thinker to develop both of these capacities.[28] As such, Laske's Framework makes an enormous contribution to the ability of individuals to develop metasystemic cognition in themselves and others.

The Framework facilitates the scaffolding of a capacity to build mental models of complex phenomena. It identifies what is missing from the models of analytical/rational thinkers and what must be included to enable metasystemic cognition.

In particular, the Framework identifies four categories or Quadrants of thought forms that tend to be missing from analytical/rational thinking. These are the Context, Process, Relationship, and 'Systems-in-Transformation' Quadrants of thought forms. I will deal with each of these in turn below.

Each category contains seven thought forms, for a total of 28. Each thought form points to a particular kind of mental representation that often needs to be included in mental models of complex phenomena if they are to be adequately represented.

When a thinker sets out to model a particular complex phenomenon, the use of the thought forms alerts the thinker to the possible need to represent aspects of complex phenomena that are often

---

[28] Laske (2023) – see References

omitted, and suggests a way to represent those aspects in the thinker's mental model.

Thought forms provide the thinker with a compendium of 'movements in thought' that have proven useful in constructing mental models of the kind of complex phenomena that the metasystemic thinker is likely to wish to model. As such, the 28 thought forms are not meant to be comprehensive and exclusive of all other possibilities. Users can and should develop new thought forms where the need arises in relation to complex phenomena they are attempting to model.

It should be remembered when using the thought forms that the Context, Process, and Relationship Quadrants each deal only with partial aspects of complex reality. They should never be used in isolation. Only when they are integrated together can they model complex, dynamic phenomena adequately. The 'Systems-in-Transformation' thought forms draw the attention of the thinker to the necessity of doing this, and to how it can be achieved. In order to model dynamical systems as they evolve and transform through time, all the relevant Context, Process, and Relationship thought forms need to be combined appropriately.

I will discuss below in detail how the Quadrants and thought forms can be used to develop metasystemic cognition. This will include descriptions of each Quadrant and pointers to how they can be used. I will also discuss in some detail the use of key thought forms from each Quadrant.

For a book-length treatment of the Quadrants and all 28 thought forms, I refer you to Otto Laske's book published by Springer: 'Advanced System-Level Problem Solving Volume 3: Manual of Dialectical Thought Forms, Second Edition'[29]. A shorter overview is freely accessible online in a three-part series of articles written by Otto Laske and published in the Integral Leadership Review: 'Teaching the Dialectical Thought Form Framework'.[30] Part 2 in the series includes a 'Compact Table' of all 28 thought forms.

---

[29] Laske (2023) – see References
[30] Laske (2017) – see References

I will now outline the Quadrants and the Thought Form Framework. Then in the second half of this chapter, I will identify how you can use the Quadrants and Framework in practices and exercises that will assist you to scaffold metasystemic cognition.

My outline of the Quadrants and thought forms focuses on how they can answer key questions that face individuals who want to self-scaffold metasystemic cognition: What specifically tends to be left out of the mechanistic models produced by analytical/rational thinking? How might the processes that tend to be left out be instead represented and incorporated into their models?

We will begin with the Context Quadrant and thought forms. This Quadrant draws the attention of the thinker to the inability of analytical/rational thinking to model and take into account appropriately the Context in which phenomena are embedded. It alerts the thinker to the need to include in their models appropriate representations of Context. The Context thought forms assist this by identifying patterns and structures that often need to be included in models if they are to represent the impact of context adequately.

This limitation of analytical/rational thinking stems in large part from the fact that mechanistic, reductionist thinking is quickly overwhelmed as the number of components that have a causative impact on phenomena increases. The greater the number of components, the more difficult it is to produce an analysable model that can be thought through. Due to combinatorial explosion, this problem tends to increase exponentially as the number of components increase.

Consequently, analytical/rational thinking tends to focus on phenomena that can be understood successfully when they are considered in isolation from the rest of reality, i.e., by ignoring the circumstances in which they are embedded. Doing so enables a major source of unanalysable complexity to be excluded from its models.

However, the downside is that this produces a major limitation in the ability of analytical/rational thinking to model phenomena effectively in circumstances where context does matter. It means that analytical/rational thinking can only deal effectively with phenomena that are not influenced significantly by the complex phenomena in which all parts of reality are embedded.

In other words, analytical/rational thinking can understand only phenomena that are largely uninfluenced by the context or bigger picture in which they are entrenched. It is no accident that the great successes of First Enlightenment science came from the discovery of physical laws that seemed to hold true irrespective of the context. For example, the laws of gravitation appear to apply anywhere in the universe, even though the context in which relevant events occur might differ significantly—the laws operate the same on the surface of the Earth as they do in space.

Furthermore, the machines produced by First Enlightenment science and technology tend to be designed to operate context-independently—they function the same irrespective of the context in which the machines are embedded.

Finally, analytical/rational thinking and the science it underpins have had little success in developing a science of phenomena that cannot be properly understood without considering their context. For example, in the case of the phenomena studied by the humanities, the impact of the context in which individuals and groups develop cannot generally be ignored. This has contributed to the failure of science to make much headway in the humanities.

Much of physics' failure to understand quantum phenomena arises where context matters: For example, whether an electron acts like a particle or a wave depends on the context. Furthermore, the 'measurement problem' arises from the fact that measurements are typically produced by a larger-scale device that imposes a change in context. The fact that context makes a significant difference to phenomena seems weird and surprising only to analytical/rational science.

If analytical/rational thinkers are to overcome this limitation, they will need to shift their attention to what is largely ignored in their thinking. They will have to give attention to the context in which phenomena are embedded and develop the ability to model its impact on the phenomena that they are investigating. The Context Quadrant alerts the thinker to the need to find ways to represent Context appropriately in their models.

However, when analytical/rational thinkers begin to give attention to modelling the relevant Context, their attempts will often fail. This is because they will tend to use only analytical/rational thinking to develop their extended models. If the context is complex and dynamic, as it often is, they will not be able to model its impact on the phenomena in question using only analytical/rational thinking. They will need to incorporate into their models mental representations of complex processes and relationships that cannot be defined in words and propositions. This is why the previous chapter emphasised the need to access 'right-brain' resources if the thinker is to model complexity effectively.

The Context Quadrant and associated thought forms assist the developing metasystemic thinker to include Context appropriately in their models. The thought forms achieve this by identifying particular patterns that are often found to be operative when Context is important in shaping complex phenomena. As such, the various thought forms draw the attention of the thinker to recurring forms of Context that are useful in understanding its causal impacts, and that will often need to be incorporated into their models.

In particular, the Context Quadrant draws the attention of the thinker to the fact that much of reality is structured in the form of a multi-level hierarchy. Reality tends to be stratified, with higher levels in the hierarchy forming the context for lower levels. There is always a bigger picture.

The context tends to constrain and control lower levels. As such, many complex phenomena cannot be understood and modelled effectively unless the impacts of relevant contextual layers are considered.

The Context Quadrant also reflects the fact that most organised wholes are comprised of multiple levels that constrain the behaviour of lower levels, and that are themselves embedded in larger systems and contexts.

A deceptively simple example is provided by a gorilla that lives within the context provided by its natural forest environment. It is likely to behave very differently from an identical gorilla that lives in the context provided by a zoo. A model of gorilla behaviour that ignores context and focuses only on the gorilla itself and the internal determinants of its behaviour will fail to account for these differences. Models that attempt to reduce organisms to the behaviour of their parts and that ignore external contexts will often fail to explain their behaviour.

I will now consider the use of two of the most significant thought forms that fall within Otto Laske's Context category. These are TF#8, which recognises that many complex phenomena are contextualised by a larger whole (in particular, this thought form focuses more on the impact of context on parts); and TF#9, which recognises that in these architectures, the whole often subordinates the parts in order to maintain the whole (this thought form focuses more on the functioning and persistence of wholes).

These two thought forms identify phenomena where ignoring the context will produce inadequate models that fail to predict how the phenomena unfold over time. This enables the thought forms to be used by a thinker to assist in identifying when and how the context in which phenomena unfold needs to be represented in their models.

Consideration of these thought forms can alert the thinker to the need to move attention out of embeddedness in analytical/rational thinking that is blind to context.

More specifically, TF#8 alerts thinkers to the fact that the behaviour of many entities/parts cannot be understood and modelled properly without considering the context in which they are embedded.

This is particularly the case when the entities and the context that influences them constitute a whole that continues to exist as an ongoing system.

Of course, such 'wholes' will also be embedded in a broader context/environment, and their behaviour can be influenced by this context. The behaviour of systems, particularly complex adaptive systems, cannot be accounted for by considering only the behaviour and interactions between the entities/parts that constitute them. Analytical/rational thinking and the mainstream science that it underpins tend to fail in these circumstances.

Numerous examples can be given of where TF#8 is relevant, and where it needs to be utilized if complex phenomena are to be modelled adequately. These include: genetically-identical twins can exhibit significantly different personality traits if they are brought up in different families; they will also tend to differ if brought up in different cultures (if an individual brought up as a born-again Christian in the United states was instead raised in a Hindu family within an Indian Hindu culture, it would be extremely unlikely that the individual would choose to embrace fundamentalist Christianity); if you were raised in a sufficiently different family or cultural context, the person who you currently experience yourself to be would not exist; the environment in which individuals develop can provide them with traumas and contradictions that propel further adaptation and development; the decisions and conscious choices that individuals make will be influenced by the stratified hierarchy of social systems in which they are embedded (e.g., family, local community, school, local government, city, region, nation-state, and so on); a full understanding of why humans and other organisms have the characteristics they do must take into account the environment/context that has shaped their adaptation/evolution (again, it is often the case that the stratified nature of the environment which comprises a hierarchy of constraints, will need to be taken into account); as I have discussed earlier in the book is some detail, the behaviour of individuals and groups is highly dependent on the nature of any management/governance that constitutes the context that influences them—change the management and you change the behaviour of the entities that are governed by the management (Evolutionary Management Theory provides common examples of circumstances in which TF#8 plays a significant role and cannot be ignored).

TF#8 alerts the aspiring metasystemic thinker to the need to take account of the kinds of contextual patterns embodied in these examples, where this is useful and necessary to model particular complex phenomena.

TF#9 focuses on the fact that within complex wholes, particularly within living systems at all levels, the context provided by the whole for its parts often must be taken into account—the context can play a significant role in maintaining and perpetuating the system as a whole, and adapting it through time. If thinkers ignore this, they will fail to be able to account for the behaviour of the system as an entirety. As such, in comparison with TF#8, TF#9 focuses more on the role of context within wholes and its function in shaping and maintaining the behaviour of the whole. In contrast, TF#8 focuses more on the impact of context on the behaviour of entities/parts.

Evolutionary Management Theory provides many examples of circumstances in which a thinker must give attention to TF#9 if their models are to represent common complex phenomena adequately. As discussed in detail earlier in the book, management/context can intervene in a group/society/system in ways that ensure that it functions effectively as a coordinated and coherent whole. Management/context can reward cooperation that enhances the functionality of the society/system, punish free-riders that undermine it, align the interests of the parts with those of the whole, ensure that parts adapt in ways that contribute to the effective adaptation of the society/system (including by implementing cybernetic feedback loops and control systems), and so on.

As we have seen, powerful contexts can also exploit groups and societies, rather than organise them into cooperative, evolvable wholes. In this common pattern, unconstrained management tends to organise a group or society so that it serves the interests of power, not those of the group as a whole.

TF#9 alerts thinkers to the need to consider the inclusion of Context in the forms identified by Evolutionary Management Theory in their models of complex reality. This is particularly the case where the focus is on the role of Context in organising a functional whole.

It is worth noting here that the Context Quadrant of thought forms focuses on the importance of 'vertical relationships' within complex phenomena. Vertical relationships include relationships between parts and wholes and between wholes and their wider environment.

Often, these vertical relationships involve instances in which higher levels have the capacity to constrain parts at lower levels. The ability to constrain is the ability to influence without being influenced in return. This is the essence of power.

As discussed previously, the long evolution of life on Earth provides many examples of management using its power to organise lower-level entities into larger-scale cooperative organisations. This potentiality has been very significant in shaping living processes. It is responsible for the emergence of human societies, including kin groups, tribal societies, kingdoms, and now human organisations that span the globe.

We will see below that horizontal relationships (as opposed to the vertical relationships that are considered in the Context Quadrant) have also been significant in the evolution of life. They are considered in the Relationship Quadrant.

A further example of the importance of TF#9 arises in psychological development, including cognitive development. Vertical psychological development provides a paradigmatic example of the emergence of a new, higher-level subject that organises lower-level processes into a new, functioning whole. The new, higher-level subject constitutes a powerful context that is capable of managing the lower, pre-existing levels.

Robert Kegan's apt description of how vertical development arises captures the essence of this kind of emergence: as we have seen, he suggests that vertical development occurs when what was part of the subject at one level becomes object to a new, higher-level subject. These kinds of vertical, psychological transitions occur repeatedly throughout the development of individuals and the evolution of humanity.

The inability of analytical/rational science to represent Context adequately in its models has largely restricted its discoveries to circumstances in which Context does not make a significant difference and can be ignored. Unfortunately, most of reality cannot be modelled successfully if Context is ignored.

Where complex context matters, mainstream science is relegated to the role of an impotent bystander. This is the case for most of the phenomena studied by the humanities. It is also the case even in the 'hard' sciences when science is forced to grapple with Context head-on. For example, in quantum mechanics where context (measurement) seems to collapse the wave function, physics has failed to make much progress.

Overall, the limitations of analytical/rational thinking have meant that science has so far been able to develop useful models of only a small proportion of reality—i.e., of the low-hanging fruit. Fortunately, with the emergence of metasystemic cognition, much more can be achieved.

The second category of phenomena that is largely left out of models by analytical/rational thinking is covered by the Process Quadrant. Analytical/rational thinking tends to omit adequate representations of Process because it is largely limited to building mental models that are analysable and that can be thought through, step by logical step.

In order for a model or a phenomenon to be analysable, it has to be constituted by elements/entities/parts that are defined clearly and unambiguously, that continue to fit the relevant definitions through time, and that have attributes that do not change significantly. In particular, analytical/rational thinking has great difficulty dealing with entities that are actually processes that are continually changing. Processes of this kind will eventually change so much that they will no longer fit the original definition that applied to them, and so on, ceaselessly.

An exception exists to the lack of ability of analytical/rational thinking to represent processes in its models. The exception arises when a process can be defined clearly and when any changes that the process undergoes can also be defined precisely. When this is the case, analytical/rational thinking is able to represent the clearly-defined processes in analysable models.

The Process Quadrant draws the attention of the thinker to the fact that their analytical/rational models will tend to be unable to understand much of reality because it is changing ceaselessly and

370

incessantly. The overwhelming majority of reality is unanalysable. The Process Quadrant alerts the thinker to the need to find ways to represent these kinds of processes in their models. If they fail to do so, their models will be unable to model successfully much of dynamic, complex reality.

In order to overcome these limitations of analytical/rational thinking, the task of the thinker is to identify patterns in the incessantly-changing processes that constitute complex phenomena. The thinker must then be able to include these patterns and processes in their models wherever they have a causative impact that is relevant to the purpose of the model.

As we have seen, these patterns and processes will not be able to be defined precisely and expressed as declarative knowledge. They will not be able to be described adequately in words, and therefore cannot be included in models as propositions. Mathematization and formalization, the holy grails of analytical/rational thinking, will be ineffective.

However, humans have cognitive and social/emotional capacities that can recognise and utilize patterns in complex phenomena. The details of how these capacities undertake these functions are generally outside conscious awareness. But the fact that they can do this is unmistakable and beyond question.

For example, we can recognise a friend's face in a crowd of a hundred other people we have never met before. And when we enter a social environment, our emotional system can quickly and consciously appraise the nature of any intense interactions that are taking place. We pick up 'the vibe' almost instantaneously. We do not have to think about it or to analyse the situation. In fact, if we do attempt to use our conscious thinking mind to figure out what is going on, doing so can exclude from our awareness important information provided by our emotional system.

I have referred to these capacities previously somewhat metaphorically as 'right-brain' resources, and the previous chapter dealt in detail with how they can be accessed. They are particularly important for enabling a thinker to give attention to the Process Quadrant and to find ways to identify and incorporate relevant patterns and processes into their models.

The Process thought forms help scaffold this capacity in the thinker. They draw the thinker's attention to the need to consider Process aspects of complex phenomena that cannot be adequately represented by analytical/rational thinking.

But more importantly, they identify particular patterns that often need to be taken into account if processes are to be modelled effectively. These are patterns that have been found to be prominent in many complex phenomena, and that often need to be represented in thinking if the phenomena are to be understood and manipulated successfully.

I will consider here the use of the three most significant and commonly-used Process thought forms: TF#1, TF#2 and TF#6.

TF#1 reminds the thinker that reality is characterised by unceasing movement. From this broader perspective, there is no such thing as an object that retains its identity indefinitely. Even a mountain range is not an object or thing. It is a process that comes into existence, undergoes constant modification, and eventually disappears.

Nor is a human a being that possesses fixed characteristics that are unchanging. Rather, a human is a process that is continually and incessantly changing and adapting.

Process TF#1 also alerts the thinker to the need to include in their models the features that propel this unceasing movement. It reminds the thinker that there is often 'an absence' in a process or system that drives this incessant change.

For example, cognitive development in humans is a process that tends to proceed throughout the life of individuals. As I have outlined in this book, it moves through a sequence of stages, potentially without end. At every step, this developmental process is propelled upwards by the existence of deficiencies (absences) in the cognitive capacity that operates at that level. Whatever the level, challenges are encountered that are unable to be resolved by existing cognitive capacities. To some extent, these absences will be overcome by the next step in the developmental process. However, some absences will still remain, and they will drive the developmental process onwards and upwards.

A further example of the important role of absences in driving change is in economic markets. The absence of products that meet a particular consumer demand will incentivize the development of new products that do so. In a sense, the absence will 'call them into existence'.

Process TF#2 identifies a very common pattern that drives ceaseless change in many processes and systems. It alerts the thinker to the need to look for this pattern when attempting to build models of complex phenomena and to incorporate the pattern into their models as appropriate.

This pattern was one of the first to be explicated in detail by earlier dialectical thinkers such as the German philosopher Hegel. It is based on the observation that, for example, when considering historical change within a society, changes will often emerge that are amplified because they meet the needs of some particular interests within the society. This is referred to as a 'thesis' in Hegelian dialectics. Very often, this emergence will evoke a counter-movement, the 'antithesis'. The counter-movement will often be driven by the fact that the initial movement acts against the interests of others in the society.

The conflict that results between the thesis and the antithesis will often be resolved by the emergence of a 'synthesis'. Within societies, this will often be produced by a governmental/management intervention that settles the conflict.

However, this process of change will not end there. The synthesis will never resolve all the relevant issues, and certainly not to the satisfaction of all members of society as circumstances continue to change. The synthesis will likely become a new thesis that evokes a further antithesis, and so on, producing the ceaseless change identified by the Process Quadrant.

Once a thinker amends their thinking strategies to include TF#2, they will start to see examples of this pattern everywhere. We have already discussed how psychological development tends to be propelled by absences in capacities that manifest in one form or another at every level. As well as exemplifying TF#1 and the power of absences, these developmental processes also provide clear examples of the dialectical process of change identified by TF#2: the inability to deal with a challenge at a particular level tends to evoke further attempts to do so. If these are unsuccessful, a higher-level self may be evoked to provide a synthesis. And so on. This book provides many examples from my life where the operation of dialectical dynamics has propelled my psychological development in this way.

Political events in human societies and throughout human history provide abundant examples of this dynamic pattern. Management/governance often provides the synthesis in specific instances. Economic phenomena and the functioning of corporations offer many more instances of this pattern.

The evolution of life on Earth also provides numerous examples. These include the emergence and operation of hierarchies of physiological and other regulatory systems within organisms; the emergence of new levels of management that suppress competition and increase integration by providing a new, higher-level 'synthesis'; and so on. As was the case with the Context Quadrant, Evolutionary Management Theory exemplifies the use of this and other Process thought forms.

Process TF#6 draws the thinker's attention to specific ways in which analytical/rational thinking fails to adequately represent processes that are ceaselessly changing. In particular, it critiques the propensity of analytical/rational thinking to reify phenomena that are, in fact, processes. Reification is the act of thinking of a form that is constantly changing as a thing or object that has fixed attributes. It fails to recognise that whatever exists is not a thing, it is a form that is characterised by unceasing change.

Using this thought form scaffolds the analytical/rational thinker to critique their own thinking and to see where it is limited and unable to adequately model complex dynamic reality. It alerts the thinker to the need to search for and practice the use of thought forms that are not handicapped by these limitations.

The third category of phenomena that tends to be left out of the models of analytical/rational thinkers is the Relationship Quadrant. In contrast to the Context Quadrant which deals with vertical relationships between entities and the context in which they are embedded, the Relationship Quadrant draws the attention of the thinker to horizontal relationships.

In particular, this Quadrant alerts the thinker to the need to recognize that the behaviour of entities/parts within a system cannot be understood without taking into account their relationships with other parts of the system. The Quadrant reminds the thinker that reductionism that treats differentiated parts in isolation from other parts will tend to fail to understand the system of relationships that constitute the whole.

For example, it is impossible to understand the functioning of the liver within our bodies without taking into account its relationship with other organs and processes, and without understanding how these relationships combine to constitute our body as an organised system that functions coherently.

Analytical/rational thinking is unable to represent these kinds of phenomena effectively in its models for several reasons. First, the entities that are involved in the relationships are, in fact, processes. They are ceaselessly changing. Consequently, they cannot be defined precisely, and any definition that might apply at a particular point in time will not continue to apply as the entities change incessantly.

Furthermore, the potential to represent these relationships in analysable models is impeded by their quantity and complexity. In any complex system, the number and diversity of evolving relationships that need to be included in a model are too numerous to be thought through in a step-by-step fashion.

Just as it did in relation to Context, analytical/rational thinking tries to cope with this unanalysable complexity by ignoring it. It tends to treat entities/processes in isolation, leaving out both the vertical and horizontal relationships in which they participate. But this works only for those limited parts of reality that can be approximated by analytical/rational models that ignore these relationships.

Analysability is further impeded by the fact that sequences of relationships within systems often feed back upon themselves, constituting circular causality that cannot be thought through.

But an even more significant limitation arises because relationships, including collective relationships, often coevolve. Within a typical complex system, processes/entities that participate in relationships tend to adapt to the behaviour of other processes/entities, and these adapt to those changes, and so on, endlessly. These coevolving processes/entities cannot be effectively represented by mechanistic, reductionist models constituted by linear chains of causality that can be analysed and thought through.

Within this context, Relationship TF#15 alerts the thinker to the need to consider whether the processes that are being investigated are part of an encompassing whole. Where they are, the thought form draws attention to what is lost if the processes are treated as isolated entities, ignoring their participation in relationships that help to constitute the whole.

This thought form is a useful guide for the thinker who is attempting to understand and model differentiated, coevolving processes that constitute larger systems. These include processes within societies, economic systems, political systems, governmental and managed systems, corporations, organisms, animal societies, ecosystems, and so on.

TF#15 draws the attention of the thinker to the fact that the relationships between the processes/entities that constitute a complex system have co-evolved in order to serve the functionality of the system as a whole. The nature of the entities and their behaviour can only be understood properly in the context of these co-evolved relationships and the larger functions that they serve.

TF#21 draws the attention of thinkers to the frequent need to include in their mental models the patterns of interaction between entities that constitute relationships. Of particular importance for this thought form is the identification of reciprocal interactions and how these patterns of interaction may change over time. However, it focuses more on the external interactions between processes/entities, rather than on how patterns of interaction may change the entities themselves as they co-evolve.

The nature of the patterns of interaction between the components of a complex, dynamical system is critically important for understanding its functioning as a whole. We have discussed earlier how the Context provided by the system for its constituents can manage/govern/constrain these patterns of interaction in ways that serve the system's functioning.

However, the significance of this thought form is not just for building models of systems that are organised into complex, functioning wholes by management/governance. It is also important for building models of simpler, unmanaged systems that nonetheless have the capacity to be self-producing.

Unmanaged systems can be self-producing where they are constituted by mutually-reinforcing relationships.

We discussed earlier in the book the example of collectively-autocatalytic sets of molecules. The ability of such a set to reproduce itself collectively though time results from the fact that the formation of every member of the set is catalysed by at least one other member of the set.

Reciprocal altruism is an example at the human level: alliances are maintained and reproduced because actions that help other members of the alliance tend to be reciprocated. A further example is the systems of exchange relations that produce and reproduce economic markets.

The same kinds of reciprocal relationships can also produce and reproduce organisations of corporations and organisations of Nations.

Whenever a new level of organisation has begun to emerge during the evolution of life on earth, it has commenced with the emergence of these kinds of unmanaged, self-producing systems. They are characterised by mutually-reinforcing, non-hierarchical relationships between the processes/entities that constitute them. Of course, as we have seen, the emergence of management (a new form of context) can greatly expand the evolvability of these previously unmanaged organisations, enabling them to evolve interrelationships that are far more complex in their service of the interests of the whole.

The fourth and final category of phenomena that cannot be effectively modelled by analytical/rational cognition is the 'Systems-in-Transformation' Quadrant. Broadly, this Quadrant integrates the

Context, Process, and Relationship Quadrants to model complex phenomena as they evolve and transform over time.

For each of the other Quadrants, we have described the particular aspects of complex phenomena that they point to. The Quadrants identify what the analytical/rational thinker must move their attention to if they are to build useful models. We have shown how the thought forms for each Quadrant alert the thinker to what is left out by analytical/rational thinking and what must now be included. We will now do the same for the Systems-in-Transformation Quadrant.

As emphasised earlier, each of the first three Quadrants is partial. Each covers only particular aspects of evolving complex phenomena. By themselves, they do not identify what needs to be encompassed by metasystemic cognition. Only when they are all considered together can they fully describe what must be represented in your models if they are to represent complex reality. This integration is not simple: the Quadrants must be combined dynamically.

The Systems-in-Transformation Quadrant moves the attention of the thinker to the whole. The other Quadrants refer only to aspects of these wholes.

Successful use of this Quadrant is critically important when thinkers are using their models to develop strategies designed to intervene in and manipulate complex reality in pursuit of human goals. If intelligent individuals want to strategize about how to achieve their goals, they must be informed by models that capture all relevant dynamics, including those that arise from human agency. Their models must include all relevant aspects of metasystemic reality as it evolves through time. The thinker must put together a multi-level, dynamically-evolving 'big picture'.

An example that is critical to human survival is the pressing need to develop effective strategies for overcoming human-induced climate change. For this, it is essential that we can build dynamic, multi-level, global models that capture all causally-relevant processes and systems, including possible human actions and their consequences.

The Systems-in-Transformation Quadrant and its thought forms are essential for scaffolding the kinds of thinking that can model human

378

agency in the world and identify how it can be enhanced to achieve particular goals.

We will consider two important thought forms in the Systems-in-Transformation Quadrant. The first is TF#23. It draws attention to a dynamic pattern that often needs to be included in models of complex phenomena if those models are to be effective and useful. This dynamic pattern points to the role of conflict in driving and resolving beneficial evolutionary and developmental change at the metasystemic level.

This book provides many examples of how this pattern operates in the development of individuals and in the evolution of life more generally. As is particularly clear in retrospect, the narrative of my own cognitive development identifies many instances in which my developmental progress would have stalled if it were not for traumas, setbacks, and conflicts.

My life is not unusual in this respect. This pattern is widely recognised in general theories of human development.

A further example is the competition and conflict that produce the natural selection that drives the continual transformation and evolution of living systems. Somewhat paradoxically, it is competition that propels the evolutionary trajectory toward increasing cooperation and evolvability. Effective cooperation is favoured by natural selection because it out-competes systems that are not cooperative. Without competition, the trend toward increasing integration and cooperation would not exist. Anything and everything would survive.

It is important to recognise that TF#23 is the equivalent at the metasystemic level of Process TF#2. TF#2 draws attention to the thesis-antithesis-synthesis pattern that is common at the process level. At the systems level, conflict between systems is often resolved by the emergence of a new level of management/governance/constraint that integrates the systems, producing the synthesis.

A current example is the conflict between nation-states that is driving environmental degradation and the possibility of nuclear annihilation. TF#23 draws the attention of the thinker to the tendency of this conflict to 'call forth' a synthesis that will resolve it. As was previously the case for conflicts at lower levels, this synthesis is likely to entail the emergence of a new level of governance, but this time at the global level.

TF#25 draws the attention of the thinker to the usefulness of being able to compare models of different dynamical, transforming systems. The use of this thought form requires the thinker to hold the systems side-by-side in mental space, enabling the systems to be compared and evaluated. This evaluation enables the thinker to compare, for example, the effectiveness of the systems, their usefulness, their stability, and their adaptability.

This thought form is particularly relevant when a thinker is attempting to model the evolution of complex living systems. For example, the thinker might be attempting to predict which forms of cooperative organisation will be competitively superior. A thinker who is able to use this thought form will be able to evaluate, for example, the extent to which particular managed organisations will outcompete unmanaged organisations, and to identify which particular management architectures are likely to be favoured by evolutionary competition.

TF#25 is also highly relevant in the design of human organisations of all kinds, from businesses, to corporations, to educational institutions, to governments, to societies, to nation-states, and eventually, to the building of a unified, cooperative, and highly-evolvable global entity. The ability to hold alternative systems side-by-side in mental space is essential for identifying the forms of organisation that will be most effective at achieving particular goals.

*   *   *

Of course, reading and understanding these descriptions of the Quadrants and thought forms will not itself enable you to be a metasystemic thinker. Nor will studying and understanding Otto Laske's Manual of Dialectical Thought Forms.

This is because the development of a capacity to actually build models requires the acquisition of particular mental skills. The Quadrants and thought forms help to identify these skills. But declarative knowledge alone about them will not install the relevant mental skills in you. This is the case with all procedural skills, like a golf swing, riding a bicycle, driving a car, and so on (not to mention a tennis serve, again).

Furthermore, it is not sufficient to understand in depth the limitations of analytical/rational thinking. It is relatively easy to understand the Thought Form Framework as a strong critique of mainstream science and of the linear, logical thinking that underpins it. But it is not enough.

This is demonstrated by postmodernism's failures. It is extremely effective at critiquing analytical/rational thinking and its products. But it has failed almost completely to develop a new, higher level of cognition that can succeed where analytical/rational thinking falls short.

Postmodernism (and the Green level in the framework of Spiral Dynamics) has not produced effective models of complex reality. It has not generated a new kind of science. Yes, it has developed an antithesis to analytical/rational thinking. However, it has not provided a new synthesis. Postmodernism has not generated a new level of cognition that is capable of understanding dynamic, complex phenomena.

Until you have developed a capacity to model complex reality, you have not achieved the higher levels of cognition that are the central focus of this book.

This is where the really hard work begins. The only way to develop the model-building skills identified by the Quadrants and thought forms is to practice them in building models. You will need to do so persistently and indefatigably until the soles of your feet and your fingernails all sweat.

Furthermore, the fact that the skills can never be specified precisely in words or in other forms of declarative knowledge, creates additional challenges. As is the case with building dis-embedding skills, overcoming these challenges demands serious experimentation and play using the thought forms.

This experimentation is essential. You will not know in detail beforehand the specific nature of the skills that you need to acquire. But if you are able to identify the particular outcome that each skill is intended to achieve, you will be able to develop it eventually, even if your experimentation relies mainly on trial-and-error. The feedback you obtain from this experimentation will guide your development of the requisite skills. Eventually, you will be able to develop a set of skills that enable effective mental modelling of complexity.

Fortunately, it is easier to test whether your experimentation with the use of thought forms is proceeding successfully than it is for the development of dis-embedding and associated skills. Whether particular model-building skills are effective can be evaluated by assessing whether they are able to build models that can make accurate predictions.

Although studying the Quadrants and thought forms will not, by itself, produce metasystemic cognition, the development of the Thought Form Framework was a major step forward in humanity's development of metasystemic cognition. Knowledge about the thought forms will not install the thought processes that produce metasystemic cognition. However, it will prove invaluable in helping individuals identify what they need to include in their mental models if they are to represent complex phenomena adequately. This guidance will then enable them to practice the mental skills that are necessary if they are to include what the Quadrants and thought forms are pointing to.

Using the Thought Form Framework will enable you to scaffold mental processes that overcome the limitations of analytical/rational thinking and to build mental models of dynamic, complex phenomena. The persistent practice of using the Quadrants and thought forms, aided by the dis-embedding practices, will eventually enable you to develop metasystemic cognition.

Broadly, the steps that you need to take to develop metasystemic cognition are: use the Quadrants and thought forms to guide your building of specific models; test your models against their ability to predict how particular complex phenomena unfold; identify where the predictions of your initial models failed; draw on the Thought Form Framework to identify why they failed; correct your models; draw again on the Thought Form Framework as necessary to identify where the meta-models that you used to generate your original models failed; amend your meta-models appropriately; test the amended models and meta-models; do this repeatedly and recursively, and so on, and so on.

The first step in this developmental process is to choose a complex phenomenon that you are highly motivated to explore. Ideally, this will involve complex systems in which you are embedded that are currently not producing the outcomes that you desire. These circumstances will incentivize you to build models that help you achieve outcomes that are more aligned with your goals.

This complex challenge may, for example, involve issues arising in your family or other close relationships; or in the organisation in which you work; or in the complex, evolving environments in which your work organisation is embedded; or in political systems in which you are involved; or in a complex sport that you play or that interests you, or in your hobbies; etc.

The next step in this scaffolding process is to refrain from researching the models, knowledge, and thinking that experts and others have used to strategize about similar complex challenges. Such research might assist you in developing effective models rapidly. But it will not assist you to develop metasystemic cognition. To do so, you need to practice using the Quadrants and thought forms to build your own models. Short-circuiting this process by adapting the models of others, in whole or in part, will impede your development of metasystemic cognition.

Instead, you should build your own models of the relevant complex phenomena from first principles, from the ground up, without reference to any existing body of knowledge.

It is critically important to be aware of the thought processes that you use as you construct your models. This is essential because later in the process, you will need to be able to remember the thought processes that you used, examine them and their effects, identify those that were sub-optimal and that led to failures in the model you developed, modify them as necessary, and thereby produce improved models and meta-models.

Consequently, you should embark on the model-building process consciously and deliberately. Ensure that you are as aware as possible of each step you take in your thinking. At first, this may prove to be extremely challenging. Initially, you may find it difficult to build the model consciously. You may have a very limited ability to be aware of each step that you take and to remember it. But as you practice building models from first principles, and as your dis-embedding practice helps to enhance your capacity to see your thinking as object, it will become easier and more fluid.

You should draw on the Quadrants and thought forms as you build your models from first principles. You may find it useful to begin by using the Quadrants to undertake an initial, broad assessment of what you need to give attention to in order to overcome the deficiencies in analytical/rational thinking.

At first, as you begin to familiarize yourself with the thought forms, you might find that you need to use them a bit like a look-up table. But as you practice and persevere with model building, you will find that your need to consult written materials about the thought forms will diminish. The need for detailed scaffolding will tend to fall away.

It is worth emphasising here that not all thought forms need to be used to build a mental model of a particular, complex phenomenon. This is because the purpose of model building is not to build a model that accurately reflects everything about a phenomenon as it unfolds through time. The only model that can include everything that comprises some part of reality is that part of reality itself. Such a complete model will not help you to understand that part of reality if you cannot understand it already. A useful map never describes everything in the territory it attempts to represent.

As mentioned earlier, the purpose of model building is to develop a model that is as simple as possible to achieve the goals of the modelling project. But not more simple. A model should therefore only use those thought forms that are necessary to enable it to predict accurately those things about a phenomenon that are of interest to the modeller.

Often, this means that particular thought forms are not relevant. This is obviously the case, for example, when the phenomena in question can be modelled adequately for the purposes of the modeller by a mechanistic, reductionist, analytical/rational model. Where it is useful, a simple analytical/rational model is sufficient. It is not valid to criticize a particular model just because it is an analytical/rational model and therefore leaves out much of the phenomena in question.

You can develop the skills needed to apply the thought forms by undertaking practices that accelerate the internalization of the thought forms and that train the mental skills that they point to.

The most significant of these practices requires you to visualize the patterns that each Quadrant and Thought Form identifies and that you may need to include in the models you are constructing. In particular, you should develop mental images that represent each of the patterns that are referred to by each Quadrant and Thought Form. Often, these visualizations will need to be dynamic, capturing the fact that most of the thought forms are pointing to patterns that are changing continuously.

Once you develop an appropriate dynamic image(s) that reflects a particular thought form, you should then practice identifying this image in complex phenomena that you encounter regularly in the world around you.

If you do this persistently and successfully, you will soon find that many of the thought forms point to patterns that are extremely common. You may find this surprising because you have never noticed them before. In large part, this is because many of the thought forms point to dynamic patterns that are not concrete. Largely, they are abstractions from concrete reality. Consequently, they are not directly observable in the world.

For example, the common thesis-antithesis-synthesis pattern cannot be directly observed in concrete reality. In order to 'see' this pattern, you need to be able to imagine the processes and relationships that produce the relevant concrete aspects of reality that you can observe directly.

You can understand and 'see' the reality and significance of these patterns only once you have developed mental models that include them. Visualizing the relevant dynamical patterns and actively seeking to observe them in real-world phenomena assists significantly in developing this capacity.

This visualization of the Quadrants and thought forms can also assist you to avoid a common trap encountered by analytical/rational thinkers when they attempt to build models of complex phenomena. When they use the Quadrants and thought forms to identify what is left out by their current thinking, and when they give attention to including representations of this in their models, they have a strong propensity to use analytical/rational thinking to do so.

Consequently, when they attempt to represent aspects of Context, Process, and Relationship in their models, they will build analytical/rational representations of them. They will tend to attempt to construct reductionist, mechanistic, and analysable representations of these aspects. They will re-embed in analytical/rational thinking.

This will not produce useful models of complex dynamical systems as they evolve and transform through time. To develop such models, the thinker needs to maintain dis-embeddedness from analytical/rational thinking as they go about building their metasystemic models.

This does not mean that you should refrain from using analytical/rational thinking at all. As for all vertical development, metasystemic cognition transcends lower levels of cognition, but also includes them. Metasystemic thinkers continue to use analytical/rational thinking, but only by choice, where it is effective. They always use it within the wider context of metasystemic cognition. Their higher cognition manages the lower levels of thinking and integrates them into metasystemic cognition.

Furthermore, the mental skills associated with analytical/rational cognition must first be mastered before they can be transcended and included in the development of metasystemic cognition.

This is a particular difficulty for many individuals who are at the Postmodern or Green stage. I am referring here to individuals who have embraced a strong critique of analytical/rational thinking without having first mastered analytical/rational thinking. They tend to be cognitive-light, right-brain dominant rather than left-brain. Under the Myers-Briggs classification system, they tend to be 'feeling' rather than 'thinking' orientated. They will not be able to develop metasystemic cognition without first having gone back to master analytical/rational cognition.

Visualizing the dynamic patterns identified by the Quadrants and thought forms will assist the developing thinker to avoid the trap of falling back into analytical/rational thinking. By focusing on patterns rather than on analytical representations, this approach tends to engage right-brain resources. Visualization and pattern recognition are predominantly right-brain capacities. Accessing these right-brain resources will assist you to dis-embed from the dominance of analytical, left-brain thinking, and help maintain this dis-embeddedness.

This will be further assisted by the practice of searching systematically for instances of particular thought forms in the complex circumstances and challenges that you encounter. As well as entrenching the visualized patterns in your mind, this practice will change and adapt the visualizations, and make them more useful to you. If done persistently, this practice will significantly enhance your development of metasystemic cognition.

When practicing the use of the Quadrants and thought forms in model building, it is also useful to remember that no list of thought forms, including the list in Otto Laske's Manual of Thought Forms, covers all useful possibilities. When attempting to build models of complex phenomena, always be on the lookout for dynamical patterns that can be included usefully in the models, but that are not covered by the existing lists of thought forms.

I have found that scientific approaches to understanding complex phenomena have identified some useful patterns that are not included in the Manual of Thought Forms. These include patterns identified by fields such as cybernetics, general systems theory, and the science of complex adaptive systems.

This is understandable, given that existing attempts to identify relevant thought forms have all originated in the humanities. The need for metasystemic cognition has been greatest in fields covered by the humanities. The humanities were unable to follow the lead of mainstream science and avoid complexity.

Although mainstream science has had little success in developing a science of evolving complexity, research at the fringes of science has made some limited progress. I incorporated some of the key patterns identified by this research in my discussion of the Quadrants and thought forms earlier in this chapter.

In summary, existing attempts to identify relevant thought forms should not be taken as complete. As you practice the development of metasystemic cognition, always be on the lookout for new patterns that can be usefully included in your model building. This is particularly the case if your interest is in modelling complex phenomena that are different from those investigated by the originators of particular formulations of the Quadrants and thought forms.

It is also important to keep in mind that the goal of these practices for developing metasystemic cognition is to become self-scaffolding. This necessitates becoming less and less reliant on external scaffolds such as declarative descriptions of the Quadrants and thought forms. Your objective should be to develop whatever capacities and knowledge you need to become self-scaffolding.

The conscious pursuit of this goal is particularly necessary if you are being assisted in your development of metasystemic cognition by a cognitive coach. Initially, the coach will help to scaffold your cognitive development by drawing your attention to what is missing from your current thinking and model building. The coach will go on to draw your attention to the thought forms that identify what you will need to include in your model building in order to correct the absences and deficiencies.

A good coach will use what Otto Laske refers to as 'mind openers' to move your attention to what is missing and what needs to be included. Laske emphasises that the Quadrants and thought forms are powerful tools for challenging and opening minds. When particular thought forms are missing from a thinker's model building, coaches can use their knowledge of the thought forms to generate questions for the thinker that will move their attention to what is absent and should be included. These questions are 'mind openers'.

We will consider examples of mind openers for several of the thought forms I discussed earlier. I will consider one from each Quadrant:

For Context TF#8, a mind-opening question for a thinker might be: "To what extent might changing the context/'bigger picture' in which the individual in question is embedded change the individual's behaviour? What kind of changes in the context might produce the behavioural changes that you are pursuing?"

An example for Process TF#2 that might be relevant to a particular thinker is: "Is there a countertendency that is arising in relation to the emerging tendency that you have referred to? What kinds of processes may emerge that might resolve these conflicting tendencies?"

For Relationship TF#15, an example might be: "What will you miss if you consider each of these processes/entities in isolation? What new insights might you have if you begin to see the relationships between them and the coevolution of these relationships? To what extent is the nature of these relationships understandable when you consider the possibility that the processes/entities are embedded in a larger whole?"

For Systems-in-Transformation TF#23, a mind-opening question asked of a thinker might be: "Could this conflict that you are experiencing eventually have positive impacts? What processes might produce this?"

The effective use of mind-opening questions will scaffold the thinker to identify and examine instances in which their thinking has failed to include dynamic patterns identified by the thought forms. The coach can further scaffold the ability of thinkers to include these patterns in their model building by leading the thinker to use the kinds of practices I have identified earlier. These include visualizing the thought forms as dynamic patterns, and actively looking for instances of the existence of these patterns in complex phenomena that they encounter.

However, if you are being coached, it is important to keep foremost in mind that your goal is to develop the capacities needed for self-scaffolding as soon as possible. This will help put you firmly on the path to recursively self-improving your own cognition indefinitely.

In an important sense, this involves internalising in your own mind the coaching process that I have outlined. Rather than the coach examining your existing thought processes from the outside and identifying their limitations, you should work on developing the capacity to do this yourself. This will entail replacing external scaffolding with internal scaffolding.

As I emphasised earlier, to achieve this capacity in full, you will need to develop the ability to see your own thinking and feeling as object. This will scaffold the emergence of a higher self that then needs to be equipped with the ability to scaffold its own cognitive capacities.

As I have outlined, this scaffolding of higher cognition can be facilitated by a number of practices. These include: The conscious and deliberate building of complex models from first principles; the use of the Quadrants and thought forms to assist this; visualizing the dynamic patterns that they identify; entrenching and habitualising their use; employing mind openers to move your own attention to where it is needed; and using the additional practices that I will identify next.

Broadly, the goal should be analogous to what should be the ultimate goal of extended psychotherapy: to enable the subject of the therapy to apply to themselves the therapeutical tools used by the therapist. Achieved successfully, this will enable the subject to self-scaffold their own therapeutic interventions indefinitely, as needed. Ideally, the goal of both the cognitive coach and the psycho-therapist should be to make themselves redundant as soon as possible. The subject should make every effort to assist them in achieving this goal.

However, your first attempts to develop models of complex phenomena from first principles will likely be riddled with failures. Initially, you are also likely to find it challenging to build your models consciously and be aware of the thought processes you use to construct them. This is why effortful work on yourself is essential.

But these challenges and failures will provide you with a good starting point for the next step in your recursive development of metasystemic cognition and beyond.

Once you have done your best to develop a model from first principles, it is time to test your model against some of the existing knowledge about the phenomena in question. However, you will need to refrain from testing it against all existing relevant knowledge.

The purpose of doing this is to begin to identify where your model is deficient. What does it leave out that needs to be included? What does it contain that should be included, but is inadequately represented?

Using the Quadrants and thought forms as necessary, you will need to identify where your model is deficient. You need to assess how it should be amended in order to ensure that its predictions are now consistent with the elements of existing knowledge that you have tested it against.

Of course, if you have done really well, you may find that your model makes some predictions that conflict with existing knowledge, but that this is not because your model is flawed. It is because your model is superior. For example, it may be evident that your model includes relevant elements that are missing from the models used to generate existing knowledge. As you develop your capacity for metasystemic modelling, this will occur with increasing frequency.

But this step of testing your model against some existing knowledge and amending it as necessary is not the most important step in this process of self-scaffolding.

You need to amend and modify not only the model you are developing. You also need to examine and amend your meta-models consciously. These are the models that you use to generate your models of complex dynamic phenomena.

The importance of this step in the self-scaffolding process is that it enables you to develop not just a better model, but better cognitive strategies and skills for developing models of other complex phenomena. It enhances your capacity to build complex models in general, not just for the specific phenomenon that you are investigating as part of this practice. This is obviously of critical importance for your development of metasystemic cognition and beyond.

The key meta mind-opening questions to ask yourself are: Why, in each specific instance in which the model failed, did your meta-models fail to develop an effective model? How can your meta-models and associated meta-strategies be amended so that they will no longer make such a mistake?

Again, you will need to draw on the Quadrants and thought forms in order to identify where your meta-strategies were deficient, and how they need to be amended. This may involve changing the visualizations that you use to envisage, detect, and represent particular dynamical patterns identified by the Quadrants and thought forms. Furthermore, as I have mentioned, from time to time it may be necessary to invent a new thought form that is better at capturing some aspect of a particular complex phenomenon you are attempting to model.

It is important that you examine and evaluate the thought processes embodied in your meta-models consciously and deliberately. This is essential if you are to develop the capacity to see your meta-thought processes as object, enabling you to evaluate them and amend them consciously.

Eventually, once you have repeated this practice sufficiently, your construction of models and your implementation of them will occur in a flash, with very little conscious awareness of the relevant mental processes. This is the same with all skills, no matter how complex—once they are fully learned and practiced, they tend to operate unconsciously with no need for deliberate, conscious involvement. Learning to drive a car is an example familiar to most—initially, full attention is needed, but eventually, a practiced driver can operate the vehicle effectively despite giving most of their attention to a conversation with a passenger.

However, this does not mean that the significance of the phase in which meta-models are constructed and amended should be downplayed. This phase is of critical importance to the successful self-scaffolding of higher cognition. Provided that you have built, adjusted, and applied your meta-models consciously in the way I have suggested, you will have the capacity to resurface them into conscious awareness whenever the need arises. This will enable you to evaluate and amend them consciously.

For example, when you find that a model of some complex phenomenon that you have developed fails in some respect, you will be able to bring back into conscious awareness the thought processes that you used to generate the model. This will enable you to identify where the thought processes failed and how they need to be changed to overcome the deficiency.

This will then enable you to give attention to a further important step. You will be able to identify whether any changes in your meta-models necessitate changes in other models that were generated using the meta-models. Making any necessary amendments to these will provide you with practice in the use of the improved meta-models. This will help to entrench them. Before long, they will be able to function fluidly and largely automatically, with little conscious involvement.

Once you have tested your models and meta-models against some of the relevant body of existing knowledge and amended them as necessary, you should repeat this process using more of the existing body of knowledge. This may include discussing your model and its implications with relevant friends, colleagues, and collaborators.

When this further testing identifies an additional deficiency in your model, you will need to repeat the process of amending and refining both your model and meta-models, including using the meta-models to identify the need to amend other models and the meta-models themselves.

Then, you will need to repeat the process, again and again, until the model passes all tests that you can reasonably make. But, over the years, you will continually be open to finding incompatibilities between the model and any new evidence about the relevant complex phenomenon.

You will actively look for any opportunity to test your models and meta-models and to obtain feedback about their effectiveness. As well as testing them against new evidence, you will take up any opportunity to discuss your meta-models and associated strategies with others who are attempting to consciously and recursively enhance their own cognitive development. Furthermore, you will read any book and consider any other resource that attempts to identify strategies for recursive self-improvement.

You will celebrate whenever you discover that your model is deficient in its ability to predict how the relevant complex phenomenon will unfold. This is because you will have encountered a new opportunity to improve not just this model, but also the whole hierarchy of meta-models that were used to generate it. The discovery of deficiencies will enable you to take further steps upward in your cognitive development.

As you improve your modelling capacity recursively in this way, you will find that you will be able to explain events that, from the perspective of analytical/rational thinking, appear to be entirely unrelated and separate. Increasingly, you will have the intellectually exhilarating experience of dots being joined, things falling into place, and events suddenly making sense in a much bigger scheme of things.

*   *   *

As humanity begins to transition to metasystemic cognition, more and more resources and external scaffolds will become available. These will assist individuals to become self-scaffolding and will support their acquisition of higher cognition. As I have mentioned earlier, this will generate a Second Enlightenment and a new kind of science.

Our economic systems will turbocharge the spread of metasystemic cognition. Any corporations whose senior executives develop metasystemic cognition will have a major competitive advantage over corporations whose executives are at the analytical/rational level.

A corporation powered by metasystemic cognition will be able to strategize more effectively when dealing with the complex, ever-evolving systems in which it is embedded. It will be far superior at predicting and anticipating changes in its economic, political, customer, and competitive environments. It will see and adapt successfully to challenges that analytical/rational corporations are blind to.

The ability of metasystemic cognition to produce overwhelming competitive advantages in economic markets will massively incentivize its acquisition and spread.

If you build a better mousetrap, the world will beat a path to your door. With little effort on the part of the inventor, the mousetrap will become widely available. Market forces will incentivize its production.

However, if you develop methods to produce a higher level of cognition that enables the building of not just a better mousetrap but all kinds of new complex inventions, the world will beat your door down. Once the benefits of higher cognition become evident, a massive new industry directed at developing and spreading metasystemic cognition will emerge overnight.

Of course, the spread of metasystemic cognition will inevitably subvert and irrevocably change the way capitalism operates currently. Increasingly, humanity will see that our societies and systems of governance will need to be reorganised. This is necessary to ensure that our economic, political, and other systems no longer self-organise environmental degradation and the threat of nuclear annihilation.

The new system of global governance that emerges will fundamentally change the behaviour of corporations and other businesses: It will align their interests with those of humanity as a whole. Corporations and individuals may still be motivated by self-interest, but the only way they will be able to satisfy their interests will be by benefitting humanity and the planet.

However, long before the spread of metasystemic cognition reaches the critical mass needed to change society, it will begin to subvert our current dysfunctional economic and political systems. When senior executives in corporations and governments develop metasystemic cognition, their belief in the value of their work will be seriously undermined. Increasingly, they will see their work and their life in the context of a much larger picture. Eventually, they will understand and embrace the biggest picture of all, the evolutionary worldview.

Within this bigger picture, their previous work and career will seem meaningless and trivial. They will see that, for example, becoming the CEO of an international corporation that makes and sells sugary drinks, is not a successful life in a wider evolutionary context. Nor is spending the money they earn from their work, no matter how much they make. Such an existence amounts to little more than spending their life masturbating Stone Age desires.

Those who develop metasystemic cognition will no longer believe in current life. They will dis-embed from the collective trance that dominates early 21st-century human existence. Although they may continue in their positions as senior executives of major corporations and pretend to pursue company goals, they will instead march to the beat of a different drummer—an evolutionary one. Propelled by their metasystemic awareness, they will become evolutionary activists. Their over-arching goal will be to do what they can to advance the evolutionary process on Earth.

Of course, as an evolutionary activist myself, the purpose of this book is to advance evolution on Earth. The book is my conscious and intentional attempt to contribute to the actualization of the major evolutionary transitions that humanity must negotiate successfully if life on earth is to contribute positively to the future evolution of life in the Universe.

Among these transitions, the priority is to develop metasystemic cognition and become self-evolving. Both of these are critically important enabling capacities. They will enable humanity to identify and implement pro-evolutionary strategies. At this perilous phase in our evolution, foremost amongst these is the imperative to establish a unified, cooperative, and highly-evolvable global entity.

References

Bejleri, E., Berberi, A., & Memaj, G., (2022). Decision Making in uncertainty and risky environment, Journal of Management Information and Decision Sciences, 25, 1-11.

Blackstone, J., (2018). Trauma and the Unbound Body: The healing power of fundamental consciousness. Sounds True: Boulder, CO.

Bostrom, N., (2003). Are you living in a computer simulation? Philosophical Quarterly 57, 243-255.

--------- (2014). Superintelligence: Paths, dangers, strategies. Oxford University Press, Oxford.

Kegan, R., (1982). The evolving self: problems and process in human development. Harvard University Press, Cambridge, MA.

--------- (1994). In over our heads: The mental demands of modern life. Harvard University Press, Cambridge, MA.

Laske, Otto, (2017). A New Approach to Dialog: Teaching the Dialectical Thought Form Framework, Integral Leadership Review.

--------- (2023). Advanced Systems-Level Problem Solving, Volume 3 – Manual of Dialectical Thought Forms, 2nd Edition. Springer Nature, Switzerland.

McGilchrist, I., (2009). The Master and his Emissary: the divided brain and the making of the Western world. Yale University Press, New Haven and London.

--------- (2021). The Matter with Things. Perspectiva Press.

Metzinger, T., (2003) Being no one: the self-model theory of subjectivity. MIT Press, Cambridge, MA.

Ouspensky, P. D., (1949). In Search of the Miraculous – Fragments of an Unknown Teaching. Harcourt Brace, New York.

Roddam, M., (2014), The Program – The Development of Meta-Systemic Cognition, Integral Leadership Review.

Ross, S., (2009). Step into the Service and Challenge of Dialectical Thinking: A Brief Review of Otto Laske's Manual of Dialectical Thought Forms, Integral Leadership Review.

References

Ruse, M., (1996). Monad to man - the concept of progress in
   evolutionary biology, Harvard University Press, Cambridge,
   MA.
Stewart, John. E., (1993). The maintenance of sex, Evolutionary
   Theory, 10, 195-202.
--------- (1995). Metaevolution, Journal of Social and Evolutionary
   Systems, 18, 113-114.
--------- (1997a). The evolution of genetic cognition, Journal of Social
   and Evolutionary Systems, 20, 53-73.
--------- (1997b). Evolutionary transitions and artificial life, Artificial
   Life 3, 101-120.
--------- (1997c). Evolutionary Progress, Journal of Social and
   Evolutionary Systems, 20, 335-62.
--------- (2000). Evolution's Arrow: the direction of evolution and the
   future of humanity. Chapman Press, Canberra.
--------- (2001). Future psychological evolution, Journal of Dynamical
   Psychology.
--------- (2007). The future evolution of consciousness, Journal of
   Consciousness Studies, 14, 58-92.
--------- (2010). The meaning of life in a developing universe,
   Foundations of Science, 15, 395-409.
-------- (2014). The direction of evolution: The rise of cooperative
   organisation, BioSystems, 123: 27-36.
--------- (2016). Review of Dialectical Thinking for Integral Leaders: A
   Primer, by Otto Laske. Integral Leadership Review.
--------- (2017). Enlightenment and the Evolution of the Material World.
   Spanda Journal, Vol VII, 1, 107-114.
--------- (2018). Evolutionary Possibilities: Can a society be constrained
   so that 'The Good' self-organizes? World Futures 74, 1-35.
--------- (2019a). The Origins of Life: The Managed-Metabolism
   Hypothesis, Foundations of Science, 24, 171–195.
--------- (2019b). The Trajectory of Evolution and Its Implications for
   Humanity, Journal of Big History 3, 141 – 155.
--------- (2020), Towards a general theory of the major cooperative
   evolutionary transitions, BioSystems, 198: 104237.

References

\-\-\-\-\-\-\-\-\- (2022). The Evolution and Development of Consciousness: The Subject-Object Emergence Hypothesis, Biosystems, 217: 104687.

Index

INDEX

INDEX

INDEX

INDEX

www.ingramcontent.com/pod-product-compliance
Lightning Source LLC
Chambersburg PA
CBHW051709020426
42333CB00014B/902